GAYS IN THE MILITARY

THE MORAL AND STRATEGIC CRISIS

Dr. George Grant, General Editor

Cal Thomas
Dr. D. James Kennedy
Lt. Col. John A. Eidsmoe, USAFR
Marlin Maddoux
Dr. George Grant
Brig. Gen. Richard Abel, USAF (Ret.)
Howard Phillips
Dr. Paul Cameron
Mark Horne
Don Feder
Peter LaBarbera
Gary DeMar
Lt. Cmdr. Gerry Carroll, USN (Ret.)

LEGACY COMMUNICATIONS
Franklin, Tennessee

Published April 1993. First Edition.
Printed in the United States of America.
97 96 95 94 93 8 7 6 5 4 3 2

Distributed to the trade through: Adroit Press, P.O. Box 680365, Franklin, Tennessee 37068

Unless otherwise noted, all Scripture quotations are from the King James Version of the Bible.

Legacy Communications
P.O. Box 680365
Franklin, Tennessee 37068-0365

ISBN 1-880692-04-X

CONTENTS

PUBLISHER'S FOREWORD

No single issue—other than perhaps abortion—has divided our nation like the issue of homosexuals serving in the military. By all appearances, this is a moral, political, and strategic powder keg which could erupt into a kind of social anarchy unparalleled in recent decades. As you will see in these collected essays, radical homosexual activists have made it clear that they will take whatever steps necessary to achieve their objective of universal acceptance of "equal rights" for their "alternative life style."

What we are witnessing is nothing less than a revolution. At the core of this revolution is a rigid doctrine which declares that the foundations of our nation's experiment in liberty are actually intolerant, unhealthy, and abnormal. The credo of this revolution—as has been the case with virtually all revolutions throughout history—is "the ends justifies the means."

Lifting the ban on homosexuals in the military is clear evidence of this. Though most of the partisans of the modern social revolution have always been at best ambivalent toward the armed forces community in the past, they have now redirected their attentions and affections—as a convenient means to their ultimate end.

The purpose of publishing this small paperback edition is to educate and alert thousands of concerned Americans to the grave implications of lifting the ban on homosexuals serving in our armed forces—not just to the military but to the culture at large. Collected from the pages of the new quarterly journal *Caveat*—a publication of Legacy Communications dedicated to the best conservative and Christia thought in America today—each of these articles offers substantial contribution to the vitality and veracity of the on going moral and strategic debate.

Congress is now deliberating on this issue and, before th end of the year, a vote will be taken. Over the course o these few months, numerous speeches, both pro and con, wi be delivered on the House floor. Expert witnesses from bot sides, along with high priced Washington lobbyists and othe influence peddlers, will fill the corridors and back rooms These folks will have the undivided attention of our electe representatives and will certainly wield incredible pressur on them. But in the final analysis, after all the smoke ha cleared, what these representatives will look at most intentl is the mail count and telephone logs from their constitu ents—the voters who sent them to Washington in the firs place. Granted, a few of our legislators are ideologicall committed to one side or the other of this debate. But th vast majority of them—both Democrats and Republicans— are still undecided about which lever to pull when vote tim comes around.

And that is where you come in.

The people who assisted in the distribution of literall thousands of copies of this book did so because they want t insure that the United States maintains its superior and un paralleled military force and to protect our culture fron sweeping revolutionary upheaval. They know that the bes way they can do that is to get you involved.

Please read this collection of essays and articles. Lear the truth about the homosexual agenda.

Then, after you have finished reading, take appropriat action. First, write or call your representatives and let then know how you feel about this crucial issue. Second, get th word out to your friends, neighbors, co-workers, fellow church members, and community leaders. You may want to order additional copies of this book of essays—or perhaps even the more complete *Caveat* edition from which these essays were collected—to give away in the short amount of time we have left before a final decision is made in Washington.

Whether the modern revolution of the homosexual extremists succeeds or fails depends almost entirely on the grassroot efforts of ordinary people like you and me.

Please do your part to preserve the health and security of our military forces as well as the stability and integrity of our national culture.

DAVID W. DUNHAM
Publisher

"Moral elements are among the most important in war. It is a paltry philosophy if, in the old fashioned way, one lays down rules and principles in total disregard of moral values."

CARL VON CLAUSEWITZ

"An imbecile habit has arisen in modern controversy of saying that such and such a creed can be held in one age but cannot be held in another. Some dogma, we are told, was credible in the twelfth century, but is not credible in the twentieth. You might as well say that a certain philosophy can be believed on Mondays, but cannot be believed on Tuesdays. You might as well say of a view of the cosmos that it was suitable for half-past three, but not suitable for half-past four. What a man can believe depends upon his philosophy, not upon the clock or the century."

G. K. CHESTERTON

"Skin color is a benign, non-behavioral characteristic. Sexual orientation is perhaps the most profound of human behavioral characteristics. Comparison of the two is a convenient but invalid argument."

GEN. COLIN POWELL

1

CLINTON'S PRO-GAY VISION FOR AMERICA

PETER LABARBERA

The gays in the military issue is just the tip of the iceberg. The Clinton administration is committed to pressing the agenda of homosexual activists into every arena and every sphere of American life. Peter LaBarbera is editor of the Washington-based Lambda Report, *a new monthly newsletter that monitors the homosexual agenda in American politics and culture. In this revealing essay he exposes the full extent of Clinton's commitment—and what it ultimately means for all of us.*

"I have a vision and you're part of it."

That pledge, made to homosexual activists at a fundraiser in Los Angeles, may prove to be Bill Clinton's best-kept campaign promise. The president's controversial push to open the U.S. military to admitted homosexuals is just the tip of the iceberg in fulfilling that promise.

Even as President Clinton reneges on key campaign pledges made to much larger, mainstream constituencies, homosexuals are trusting that he will make good on his commitment to their political agenda. Clearly, their multi-million dollar investment to elect him president has paid off in the early days of the new administration.

Gays reportedly contributed over $3 million to Clinton's campaign, worked diligently for his election, and voted for him in overwhelming numbers. Now they expect results, and fully expect to be "part of the governing coalition," as a leading gay activist put it.

"I have spent the last year of my life telling this community that our agenda will be accomplished through this administration," said David Mixner, a longtime Clinton friend and advisor who turned down a job in the Clinton administration so that he could lobby the president to fulfill his promises "in a timely manner."

Another gay Clinton benefactor, recording industry millionaire David Geffen, has pledged to spend whatever it takes for a public relations campaign to win Congressional assent to the president's goal of ending the military's ban on homosexuals.

PRO-GAY APPOINTMENTS

Bill Clinton has appointed a number of pro-homosexual political figures to his administration, including:

Defense Secretary Les Aspin. Some Capitol officials believe Aspin was picked over Senate Armed Services Committee Chairman Sam Nunn because of Nunn's outspoken opposition to lifting the ban on homosexuals in the military. Indeed, during the transition period, gay extremists staged a "kiss-in"—homosexuals kissing each other in public for shock effect—at the offices of Sen. Nunn to protest his past firing of gay staffers. Aspin strongly supports lifting the military's gay ban—a key litmus test for the homosexual lobby that poured money into Clinton's campaign.

Health and Human Services Secretary Donna Shalala. Shalala is another favorite of gay activists. Following her appointment, a headline in the homosexual *Washington Blade* newspaper called her a "strong ally." There is good reason for the euphoria. Within a month of becoming chancellor of the University of Wisconsin in 1988, Shalala went to work on codes banning "discrimination on the basis of sexual orientation" at the school. The extremist group Queer Nation held a December news conference to announce that Shalala is a lesbian, and called on her to admit so publicly. But Shalala denies the allegation.

Commerce Secretary Ron Brown. As Democratic National Chairman, Brown presided over a party that includes a Gay and Lesbian Caucus. After Clinton's election victory, Brown made a highly visible stop by the gay victory party. Photos of Brown shaking hands with exuberant gays, whose festivity was down the hall from the Democratic National Convention's official bash at the Omni Shoreham Hotel, appeared in the *Blade* over the headline, "We did it! We did it!" Earlier in 1992, Brown penned a scathing column for the homosexual magazine *The Advocate* that lauded the Democrats as the party of gay rights and excoriated Republican "bigotry" on the issue.

Housing and Urban Development Assistant Secretary Roberta Achtenberg. A militant lesbian activist and the first openly gay senior presidential political appointment, Achtenberg lobbied for the post so she could promote homosexual causes in federal housing projects. In San Francisco, she led a spiteful campaign against the Boy Scouts of America for its policies banning gay scoutmasters, not allowing girls, and for including the name of God in its pledge.

Surgeon General Joycelen Elders. As Health director for Gov. Clinton, Elders was a vocal proponent of the "save sex" (not to be confused with the liberal "*safe* sex" message!) condom message. She speaks aggressively for what she believes and can be counted on to use this key "bully pulpit" to promote various aspects of the gay agenda.

In addition to the above, Clinton has named a number of open homosexuals to lesser—yet strategic—posts in his administration. By mid-December, the *Washington Blade* reported that the president had named thirteen open gays to his transition team. And on March 12, the *Blade* reported that four more homosexuals had received jobs in the Clinton administration—including Bob Hattoy, an AIDS sufferer who addressed the Democratic convention, as associate director of White House Presidential Personnel.

"I think we're going to see some more appointments from the gay and lesbian community," said Romulo "Romy" Diaz, Jr., the newly appointed deputy chief of staff and counselor to Energy Secretary Hazel O'Leary. "There's a lot of support from the administration for the efforts of the gay and lesbian community."

Indeed, there is little doubt that President Clinton feels a strong sense of loyalty to the gay movement—both philosophically and for helping him get elected. His post-election thank-you letter to a gathering of homosexual activists was a watershed event in gay politics. Clinton's letter "to my friends at the National Gay and Lesbian Task Force"—the movement's premier political lobby—praised gay activists for their work on his campaign. The note was read by outgoing NGLTF director Urvashi Vaid at the organization's poitical conference in San Francisco, igniting prolonged cheering and applause.

SOME OF CLINTON'S PROMISES TO HOMOSEXUALS

The following are Clinton's promises to homosexual groups, as culled from the gay press and the president-elect's responses to surveys. In addition to lifting the ban on gays in the military, the president has pledged that he:

- Will issue a similar order barring "discrimination based on sexual or affectation orientation" in the federal government—effectively giving homosexuals protected "minority" status in the two-million-member federal bureaucracy;
- Will support a national "gay rights" bill. Clinton supports (with technical adjustments) the bill to amend the Civil Rights Act by adding "sexual orientation" to the list of protected classifications. The bill would make homosexuals a protected "minority," akin to racial minorities and the disabled;
- Will implement affirmative action policies for homosexuals. Clinton answered "yes" to a survey question by the Human Rights Campaign Fund (HRCF), the nation's largest gay lobby group, asking whether he had "instituted an affirmative action program to hire qualified lesbians and gay men, ethnic minorities, and women on your campaign, and will you do so as president?";
- Will support condom distribution programs;
- Will increase federal funding on AIDS research as part of a "Manhattan Project" to wage a real "war" on the

disease. (AIDS research already receives more federal monies per capita than both cancer and heart disease;

- Will appoint an "AIDS czar" to implement all the recommendations of the president's National Commission on AIDS. Among the commission's recommendations is a massive federally-funded public education campaign built around the "safe sex" message stressing condom-usage. Clinton responded "yes" when asked by HRCF if he would "support legislation and policies to create effective AIDS prevention programs that disseminate accurate, frank, and explicit information to reduce the spread of the illness." (Note: such programs favored by homosexual groups often contain graphic depictions of or allusions to gay sex acts.);
- Will oppose content restrictions for the National Endowment for the Arts, the Public Broadcasting Service, and National Public Radio;
- Will demand that the Justice Department crack down on "hate crimes" against gays and lesbians;
- Will oppose mandatory HIV testing in *all* circumstances.

GAYS "NOT LEAVING THINGS TO CHANCE"

Despite the seeming loyalty of the Clinton administration to gay causes, the *Washington Blade* reports that major gay organizations are "not leaving things to chance" and are preparing strategic plans to ensure that Clinton will keep his promises to enforce their agenda. Immediately after the election, a homosexual coalition supplied Clinton with a "presidential appointments project" of prospective gay political appointees.

According to the *Blade,* the National Gay and Lesbian Task Force has prepared a transition paper that urges Clinton to take action in several areas. In addition to some of the policies listed above, the Task Force recommended quick action on statehood for the District of Columbia (a bastion of gay political power) and "family diversity"—the movement's code word for non-married "partner" and multiple partner households to be given the same status as families.

The heavy gay input has had its logical effect. In the midst of the political firestorm over the president's proposal to allow open homosexuals in the U.S. military, Defense Secretary Aspin floated a proposal that would have ended the current policy of investigating gays in the services—but *without* a formal executive order. The proposal was jettisoned after heavy insider lobbying by gays, who insisted that Clinton sign an executive order that parallels President Truman's 1948 desegregating the military. Most Americans reject this principled analogy of gays to blacks. By backing down, Clinton showed he would rather stay true to the tiny homosexual lobby than embrace a compromise that offered him potential political gain with the mainstream Americans.

Gay media also report that the cadre of gay leaders inside the administration is preparing to push for the national homosexual "civil rights" bill. "It was never a matter of whether we should be doing this," Hattoy told the *Blade*. "It was a matter of how it should be done." Homosexual activists hope the national legislation will override anti-"gay rights" state laws like that which recently passed in Colorado—laws that gay Clinton advisor David Mixner said have the political potential for "bleeding [gays] dry." In the wake of the fractious debate over ending the military gay ban, Clinton may elect to go slowly on pushing a national "sexual orientation."

CAMPAIGNED HARD FOR GAY VOTE

During the campaign, gay support for Clinton swelled as the Democrats pursued homosexual voters with vigor. The *Blade* claims that "thousands" of homosexuals worked on the Clinton campaign. And homosexual activists point proudly to a Clinton speech in which he told gays, "I have a vision and you're part of it." Videotape copies of the speech were hot items among homosexuals across the country. The quote became the slogan of the Gay and Lesbian Inaugural Ball—officially endorsed by the Clinton-Gore Inaugural Committee—the night the new president and vice-president were sworn in.

Clinton, for his part, pleased the gay lobby when he mentioned AIDS as a top priority in his very first speech as president-elect, delivered on the steps of the state capitol in Little Rock, Arkansas on election night.

WILL CLINTON BREAK GAY PROMISES?

Bill Clinton is nothing if not politically savvy, and gay activists are fretting that he is starting to back away from their agenda—despite all he has done so far to support them. The gay press vented some of these worries after Clinton, in a meeting with House Republican leaders, reportedly indicated that he will not try to end the ban on HIV-infected immigrants after the House and Senate voted overwhelmingly against changing the policy.

The negative gay reaction to the congressional action that was immensely popular with the rest of the country points clearly to the risks of Bill Clinton's embrace of the tiny—yet demanding—gay movement. Homosexuals may comprise only 2.4 percent of U.S. electorate—as an anonymous exit poll found on election day—but they are an extremely vocal 2.4 percent!

During the election, candidate Clinton did disappoint the gay lobby on at least two points: First, he came out against homosexual marriages, and second, he opposes forcing the Boy Scouts to allow gay scoutmasters. However, both these aspects of the gay agenda could be achieved through pro-gay lawsuits filed under a national "sexual orientation" law—which homosexuals are counting on Clinton to support and for which his gay advisors are planning passage.

It seems the president will be under even greater pressure to abandon some aspects of his pro-gay "vision," since he has already ditched pledges to much larger political constituencies. The odd position in which the president finds himself is perhaps best illustrated by the following exchange between a lesbian and Clinton the day after his inauguration. Here is the conversation between the president and an apparently liberal lesbian who stood in line with other gays at the White House open house on January 21, as reported verbatim by the gay magazine *Frontiers*:

> "You can keep your tax cut if you [promise to] keep your promises to the gay community," said one woman to Clinton, shaking his hand eagerly.
>
> "You know what?" he told her with a broad grin. "You've got yourself a deal."

2

SHOULD OPEN HOMOSEXUALS SERVE?

RICHARD ABEL

The crisis over lifting the ban on homosexuals in the military is at root a moral and ethical crisis. Even so, it also involves the basic military doctrine of "necessity." In this important essay, one of our nation's most outstanding military leaders, Brig. Gen. Richard F. Abel, USAF (Ret.), dispassionately demonstrates how unit cohesion, general morale, troop discipline, and combat readiness enter into the issue.

Privacy. Freedom of association. Sexual preference. Individual rights. All these and many, many more are freedoms near and dear to every American.

They should be. They've been paid for dearly, in civil and military conflicts over the past two hundred years.

The problem is that each of them also happens to be a privilege at least partly set aside by every man and woman who voluntarily chooses to enlist or accept a commission in the armed services of the United States.

Sexual preference, for instance, is a matter of choice—clearly unlike race or gender, which are matters of birth. Sexual preference is a choice civilian Americans are free to make, but a choice set aside under current law and regulations by all those who choose military service. There are reasons for that—good and well-thought reasons rooted in military law and time-proven fundamentals that make for battlefield effectiveness. The four primary reasons are unit cohesion, health, morale, and discipline—and each are still valid today.

Unit cohesion is essential for any fighting force. In the very arduous pressure-cooker atmosphere of daily military life, trust, confidence, and mutual respect are non-negotiables. They are absolutely required—aboard ship, in the barracks, on the flight lines, and in the foxholes—where men depend on each other for their very lives as members of a mutually-supportive fighting team. All of that is undermined, if not compromised altogether, by the uncertainties, wariness, distrust, and discord open or suspected homosexuality will introduce.

Health is another undeniably vital issue in an environment where exposure to personal injury is a necessary and accepted part of daily life. Heretofore unimagined challenges will be visited on commanders, soldiers, sailors, airmen, and marines by openly homosexual servicemen and women whose chosen lifestyle exposes them and the overall unit blood supply to a much higher risk of AIDS and other potentially fatal or disabling infections.

Morale makes the difference between winning or losing, for a team or an inspired, cohesive, and successful military force. It is hard to see how unit morale will improve when people many would prefer not to live with start coming out—under court order or executive decree—in the foxholes or flight lines.

Privacy and respect are also vitally important, and must not be overlooked. Privacy is hard to come by in the military, whether in the barracks or aboard ship, where dozens of men share communal showers or toilet facilities. Privacy is also hard to come by in the field, where, again, modern conveniences are almost unheard of. And privacy is in short supply aboard submarines, where cramped living spaces require many sailors to "hot bunk," as they call it, with two or three men using the same beds on a time-share basis.

In today's all-volunteer force, male and female service personnel are accorded separate sleeping, bathing, and living accommodations. Introducing known homosexuals into the military community is nothing less than an invasion of what precious little privacy still remains for the far majority of military personnel.

Discipline—the voluntary or directed subordination of self-will to the greater good of the unit—is inescapably essential for unit cohesion and military readiness. Somewhat akin to the civilian-world principle of eminent domain, discipline is the antithesis of self-serving self-advancement, and the touchstone of combat effectiveness and military success.

One final word may be good to keep in mind: "Selflessness." It forms the bedrock supporting the very concept of military or civil service and is inseparable from the personal and corporate sacrifice which national service—in or out of uniform—is all about.

Yes, Uncle Sam is looking for more "proud, brave, loyal, and good Americans" who are willing to serve our nation. America has many needs for truly selfless, tireless servants. Those who answer the call, but who have chosen the homosexual lifestyle can, and clearly would, serve our nation best outside the military services.

3

BEATING THE LIBERALS AT THEIR OWN GAME

GERRY CARROLL

More often than not, in the heat of the fight, liberals resort to name-calling, ad-hominem, and hyperbole. Rational argument, verifiable facts, and the dumb certainties of experience seem to matter little. How do you argue with that? According to Lt. Cmdr. Gerry Carroll, USN (Ret.), you don't. Instead, you take your case directly to the people. In this trenchant essay, he shows why—and how.

I can imagine few things more destructive to the military—upon which we have depended for two hundred years for the safety and security of our way of life—than to try to integrate homosexuals into it. Forcing young servicemen and women to live in the close contact of others with a chosen lifestyle that is utterly repugnant to them will ultimately put this nation in the position of countries like France or the Netherlands, who speak loudly but carry no stick at all.

One theme that shows up over and over in discussions of this particular point is the phrase "good order and discipline." I think the best way to understand that is to think of it this way: Good order and discipline are to the military what DNA is to life. The military cannot exist without them.

What homosexuals want us to accept is that they'll come into the military and behave themselves like good little boys and girls. They won't allow their predilections for deviant behavior to interfere with the order and discipline of the units to which they are assigned. Now this is from people who average fifty sexual partners in a year as opposed to a heterosexual's twelve in a lifetime.

The average age of someone in the military is ninteen and I cannot imagine anybody whose hormones are more fired up than a nineteen year old's. (I've got one at home.)

In a military situation you live in closer contact with others than most people can believe. On a ship there are sometimes eighty men living in a space 10 feet high and about half the size of a tennis court. Everything you own is in a little locker like the ones in a high school passageway or a little lockable space under your rack. Some say the principle difference between being in prison and being in the Navy is that in the Navy you stand an excellent chance of drowning.

You don't get to go home when the workday ends. You go back to your eighty man compartment and spend all of your free time there. You do this for months at a stretch. And in living this closely you quickly learn to ignore the little benign differences in people like race or religion, because a person's color or faith does not intrude upon your life or the effectiveness of your unit. The reason a restrictive way of life like this works is that everyone *behaves* predictably. Everyone behaves to the norm. There is no room for alternative lifestyles.

Sexual tension strikes at the very core of human relationships. To heterosexuals, contact with homosexuals, however small, is repugnant. No matter how hard one tries, there is always some reluctance to come into contact with that person again. So what happens in the military? You're forced not only to be in contact with the person, but to eat, sleep, live, and work with him. You can't go home; you can't get away. (On a carrier, the farthest you can get from another person is 1,000 feet for seventy-seven days at a stretch. 1000 feet.) And by your very nature—and by his—you can never trust or comfortably work around that person again; therefore, your unit necessarily functions less effectively as a team. When the shooting starts, people are going to die. That's the bottom line: People are going to die.

That's a pretty simplistic explanation of the effect of homosexuals in the military. But there's a little more to it. We're being told that banning homosexuals from the military is discrimination. And in a sense, it is. Homosexuality is by its very nature abnormal behavior. So are bestiality, pedophilia, necrophilia, cleptomania, and serial murder. It's true. The military either prevents these people from joining or throws them out when they're discovered. And they throw them out rather gently, I think. There's really no mark on their record.

The Supreme Court's 1974 decision in the *Parker v. Levy* case affirmed that the military is "a specialized society separate from civilian society." The military *has to* exclude people who are too short, too fat, too tall, not smart enough, not literate enough, and those with infectious diseases or certain physical handicaps. Exclusion based on these traits is a perfectly sound practice. It's sensible. The traits simply make a person unable to do the job. However, they do nothing to interfere with other people's ability to do their job.

Homosexuality works exactly in the reverse.

Now what would happen if the president issued an order forcing young women in the military to share a room or a shower with young men. Well, the uproar would be historic. Everyone knows that young men would look at naked young women with something less than cold detachment--most of the sailors I know anyway. So why do proponents of lifting the ban expect us to believe that homosexuals can shower with the objects of their desires and remain unaffected. It just doesn't make sense.

Study after study has shown that the most sexually aggressive and sexually irresponsible group in the world is homosexuals. They are between twenty and fifty times more likely to contract sexually transmitted diseases than heterosexuals. Those diseases normally remove an individual from duty status for the course of treatment. Once again, we can end the effectiveness of his unit; if he's not there, he can't do his job. Also, since members of the military are a walking blood bank, the spread of those diseases to others is a certainty. In addition, 25 to 30 percent of homosexuals are alcoholics as opposed to 10 percent of the general population. In my experience, the vast majority of disciplinary problems among the sailors were attributable—at least in part—to alcohol abuse.

The problem we are going to have in keeping the ban in place is that statistics like those above and phrases like "good order and discipline" are going to be dismissed out of hand. They're problems we'll worry about sometime down the road. We're being told that the goal is to lift an unfair, discriminatory ban on homosexuals because they're just like everybody else. Well if they were just like everybody else, they wouldn't be homosexuals.

We consistently use logic and reason when we try to fight liberal ideas and programs. We should learn that logic and reason do not work. If logic and reason were universally applicable, there would be no such thing as a liberal. What we're doing here, in my opinion, is bringing a knife to a gunfight. We try to explain the dangers of lifting the ban. The liberals push the button and launch their *bias* and *bigotry* missiles at us.

You know what a bigot is, right? Someone who is winning an argument with a liberal.

We then spend so much time and effort defending ourselves against this *ad hominem* attack that we lose the argument. We have been beaten this way over and over and over again. This present battle is going to be won only if we use the liberals' tactics against them. If they push their bigot button, fine. Ignore it.

Make the case to the people that they really don't want their sons and daughters forced to live in close contact with sexual deviants. Explain that their kids will be forced to live with a group, half of whom carry communicable disease—*by choice.* And when they beat the drum that everybody should be free to choose their own style of sexual fulfillment, then simply point out that child molesting and bestiality are sexual preferences just like homosexuality. Jeffrey Dahmer, after all, was merely expressing his preferences and orientation.

I think we have to paint our arguments in the blackest terms possible. If we lose this one, I think we stand to lose a lot more besides. Personally, I could not care less about the homosexual's problem with self-esteem. I'm afraid we're about to let our moral standards decay into invisibility. If that happens, there will be no one to blame but us. The liberals aren't smart enough to see that far down the road. But we have to be.

TROOP IMMORALE

GEORGE GRANT AND MARK HORNE

In this excerpt from their most recent book-length collaboration, Legislating Immorality: The Homosexual Movement Comes Out of the Closet *George Grant and Mark Horne outline the disturbing historical background of the current flap over homosexuals in the military. No one can deny that moral breaches have been a part of military life from time to time in the past—at least, not after reading this essay. But according to Grant and Horne lifting the bar on sexual impropriety could very well transform those heretofore isolated incidents into accepted and normative behavior. It would insure that a veritable Pandora's Box will be opened—thus simultaneously undermining military discipline and putting our national security at risk.*

In 1945, a young naval recruit was sound asleep during his first night aboard ship when he was abruptly disturbed. "The awakening was sudden and panic-filled. A hand was caressing my leg, running up the inside of my thigh. A dim figure ducked away as I lashed out, kicking, swinging a fist and striking the air. There was no more sleep that night."

Kevin McCrane, now a retired businessman living in New Jersey, had been drafted into the navy at the close of World War II. In January he was assigned with four others to the *USS Warrick*, a cargo carrier. The day after McCrane was sexually harassed, the ship set sail for Honolulu. "But the excitement was gone, at least for me. At the end of a long day riding the sea's rolling swells, I took a twelve-inch, box-end wrench from the engine room and retreated to my berth. Hanging onto the wrench under my pillow, I slept."

Had this been the only incident, McCrane's voyage would have been bad enough. But it only got worse. On his fourth day at sea, he went to the *Warrick's* post office. There he was "warmly" received by the second-class petty officer in charge. Perhaps too warmly: "Grinning broadly, he stepped back from the counter, dropped his dungarees, fondled himself and made an obscene invitation. I walked away."

He went to the third-class petty officer on his watch to report the sexual harassment. "He laughed at what I told him," remembers McCrane. "He told me to watch out."

The new recruits aboard the *Warrick* soon compared notes and discovered that all of them had recently been "accosted, patted," and "propositioned." They soon learned that there was only safety in numbers: "Though we were in different divisions, we flocked together for meals, averting our eyes when one of *them* leered in our direction."

Avoiding "them" became a constant struggle for sexual survival aboard the ship:

> There were five such aggressive homosexuals that we knew of on board this ship with almost 250 men. They were all petty officers. Their actions were enough to poison the atmosphere on the *Warrick*. Meals, showers, attendance at the movies, decisions about where you went on the ship alone—all became part of a worried calculation of risk.

The consequences for failing to evade these officers and getting caught by them alone were quite severe. McCrane received news from a tearful fellow recruit that "the smallest and most vulnerable" of the recruits had been caught in the paint locker and forcefully sodomized.

At the end of the voyage the ship was given a new executive officer who summarily transferred each of the homosexual officers off his ship. This caused the crew to break out into spontaneous cheering.

EVEN IN THE MILITARY

Sadly, McCrane's awful experience was not entirely unique. Despite the historic ban on homosexuals in the military, there have been a number of homosexual rape incidents through the years in all four branches of the American armed forces. Try as it may, the military simply has been unable to maintain absolute immunity from the forces of moral disruption infecting the society at large. Though the men and women who have served our nation with courage and distinction in the military have more often than not been among the best America has to offer—paragons of virtue, discipline, and uprightness—they have sometimes also been among the worst America has to offer. Though the recruitment process has always been designed to screen out those less desirable elements, like everything else in this poor fallen world, it is a fallible process.

That process has proven to be particularly fallible when it comes to identifying and disqualifying homosexuals from military service. And it appears that the evidence is not merely anecdotal. A decade ago the *American Journal of Psychiatry* reported on a study of thirteen homosexual assaults in the military, eleven of which were carried out by more than one attacker.

For example:

One sailor was referred for a psychiatric evaluation two weeks after being physically and sexually abused when he was given a "blanket party" the first day he reported on board his ship. A blanket party consists of several men forcibly wrapping the victim in blankets so that he is unable to determine exactly who is sexually abusing him.

After that attack, the sailor was subjected to an even more humiliating and perverse assault:

One week later he was given a "greasing" by three shipmates. A greasing involves stripping the individual naked and massaging him with a thick black grease used to lubricate heavy machinery. In some cases a flexible tube is forced through the victim's anus and into his rectum. The tube is connected to a cylindrical reservoir filled with the lubricating grease. The reservoir and tube resemble a large hand-driven pump. The contents of the piston-driven reservoir are then pumped into the victim.

Unlike the blanket party, this time the victim clearly knew who his assailants were. He reported them and judicial action was taken. This did not end his torment, however. The sailor began to get threats from other shipmates that he would be physically hurt, raped again, or even thrown overboard once the ship was sent out to sea. Finally, he went to the ship physician and begged him to get him off the ship. "Get me to my dad and out of here, away from this," he pleaded. "My dad can help me."

Another sailor was also assaulted twice:

He was overpowered by three shipmates, beaten, and dragged to a secluded food storage area on the ship. Although he resisted, his attackers undid his pants and attempted anal intercourse, but he was able to escape. He was threatened if he reported the incident. He then left his duty station without authorization. This broke a one-and-a-half-year record of excellent adjustment to the Navy. He returned after a few weeks and

> was again attacked by the same three in the same area. This time he was beaten, stripped of his clothing, held down by two of the three, and raped anally by the third. He again left on unauthorized leave.

The sailor found it almost impossible to deal with the homosexual assault. His state of mind became increasingly worse.

> His service adjustment continued to deteriorate. He was frightened about returning to the ship for fear he would be assaulted again. When he did return, he carried a wrench for protection wherever he went. He was afraid to tell anyone of the assault for fear of being labeled "queer" and being discharged from the navy for homosexuality. When he did inform his superiors, no one believed him. Since the initial attack, he reported being increasingly angry and experiencing insomnia with nightmares and dreams reliving the assault.

Though certainly isolated and extreme, these incidents offer evidence of a kind of ongoing covert homosexual counter-culture within the military.

MAKING THE MILITARY SAFE FOR HOMOSEXUALITY

What is amazing about these incidents is that they all occurred while the ban on homosexuals in the military was in full force. Indeed, Dr. Peter Goyer and Dr. Henry Eddleman wrote that they became interested in studying homosexual rape in the military because they were surprised at the number of such cases:

> For the past few years we have worked in a psychiatric outpatient clinic serving a predominantly male population of active-duty Navy and Marine Corps personnel. In our work with this unique population we became aware that we were evaluating more adult male victims of male assault than had been suggested in the medical literature.

But it could have been worse. Much worse.

In fact, despite the surprising frequency of homosexual incidents that such studies have recently uncovered, they are still very much the exception rather than the rule. Dr. James Gilmore, another researcher who has studied the problem of sexual assaults in the military, has said:

> There is no doubt that in a semi-closed community like the armed forces, tragic incidents like these can and do occur—sometimes more often than we'd like to admit. But there is also no doubt that they are ameliorated in both their frequency and intensity by an enforced code of conduct. The very fact that homosexual activity within the four branches of the service has always been forced underground is testimony to the efficacy of the ban—in both inhibiting that activity and protecting innocent parties from libertine unrestraint. Without such a code, violations of persons, property, and propriety would undeniably escalate beyond bearable limits—and commanding officers would be left without appropriate disciplinary recourse. The ban is a necessary hedge of protection against any further moral or strategic erosions.

But now it appears that Bill Clinton wants to do away with that "hedge of protection" altogether. In fulfillment of his campaign pledge to lift the ban on homosexuals in the military, he announced in the first full week of his presidency plans to bring gays in the military out of the closet. Past presidents have expressed their desire to use the military to make the world safe for democracy. President Clinton apparently wants to use democracy to make the military safe for homosexuality.

Needless to say, the president's position has provoked a firestorm of protest including a deluge of irate letters, phone calls, and telegrams—more than at any other time in recent memory. According to one national survey nearly 82 percent of Americans opposed lifting the ban. Another poll determined that 97 percent of the military community was opposed to the president's plan and more than 45 percent said that they would consider leaving the military altogether if it were actually implemented.

Once the president's advisors and handlers began to realize the magnitude of the opposition arrayed against him, they suggested that he offer a "compromise," in which he promised not to lift the ban by immediate executive order, but would wait six months and then remove it only if warranted. In the meantime, he would direct the military to disregard the ban on homosexuals.

Not surprisingly, that kind of transparent dodge did not make anyone too terribly happy. Indiana's conservative senator Dan Coats declared:

> Given the president's fixed position, it appears that any compromise or negotiation on this issue is doomed before it starts. A six-month period of study when the conclusion is foregone is simply a political cover.

Meanwhile, liberals were disappointed that the president caved in—however insincerely—to political pressure. Massachusetts congressman Gerry Studds, himself an avowed homosexual, complained:

> The administration needs to decide whether or not it has the courage to lead, to decide to do right, even when the decision is unpopular or difficult. I fear for this administration should the president break his promise.

But such a fear is likely unfounded. After all, throughout his campaign for the White House, Bill Clinton not only promised to lift the ban on homosexuals in the military, he promised the passage of hate crimes legislation, preferential hiring codes, and massive increases in AIDS funding. And while he has shown absolutely no qualms about breaking his promise of a middle class tax cut, he knows that he must appease the homosexual activists if he has any hope of holding his tenuous coalition together. Besides that, the president has no desire to see—in the words of one homosexual columnist who apparently doesn't mind resorting to base stereotypes when it serves his purposes—"upwards of one million raging queers storming the White House."

DETAILS, DETAILS

The fact that Clinton's position was from the beginning ideologically driven rather than pragmatically informed means that the inherent problems involved in lifting the homosexual ban were never really taken into consideration. Such "meddlesome details" as the disposition of military housing, spousal benefits, pensions, health care, and on-base social clubs have simply been delegated to Les Aspin to take care of.

Robert Morrison, a lobbyist on Capitol Hill for the Family Research Council, has said that even Washington Democrats have complained about the nagging but neglected complications involved in lifting the ban:

> Senator Sam Nunn has raised a number of questions, such as: What are you going to do about formal military dress balls? Are you going to allow military officers to come and dance with homosexual dates?

It seems that ideological pronouncements rarely take such mundane practical concerns into account.

"The next thing you have to face," Morrison said, "is quotas for advancement, promotion, and retention." Though the military does not have an official quota system, it does have something perilously close to it. When a minority member of the armed forces is turned down for a promotion, the promotion board must be prepared to defend that decision, perhaps several times. In contrast, promotion boards generally do not have to answer to anyone if they by-pass a white male. Since lifting the ban on homosexuality has already been compared to ending discrimination against minorities, there is every reason in the world to expect homosexuals would get that same kind of preferential treatment.

Even if all these problems were resolved, Clinton would still face a huge obstacle. He originally promised to overturn the ban by Executive Order. As columnist Jeffrey Hart has argued:

> Nothing can be more fundamental to the Constitution than its first article, which defines the powers of the president and the Congress. Article 1, Section 8, says, "Congress shall have power . . . to make rules for the government and regulation of the land and naval forces." It is spelled out quite clearly. . . .Thus Congress, by law, enacted the homosexual ban. It was Clinton's delusion that he can overturn such a law "with the stroke of a pen." . . . It makes one wonder what they have been teaching at the Yale Law School.

If the president attempts to issue an Executive Order he will be flatly violating the Constitution's separation of powers. Nevertheless, not one to be hampered by such extraneous legal technicalities as the Constitution's mandated separation of powers—we've got to get rid of gridlock after all—Clinton is determined to pursue his goal of an omnisexual army. Regardless of the details.

COURTING DISASTER

One convenient way liberals have always found to get around the inconvenient details of political reality is an activist use of the courts. Thus it came as no surprise when just as Clinton seemed hopelessly mired in his quixotic crusade to lift the homosexual ban, he received a big boost from the judicial branch. Federal District Judge Terry Hatter, a Carter appointee, suddenly ruled that the long-standing—and oft court-tested—ban was actually unconstitutional. Without even a hint of complicity, White House press secretary Dee Dee Myers told reporters that she did not expect the ruling to be appealed by the Justice Department.

Tom Jipping, the Director of the Center for Law and Democracy of the Free Congress Foundation, said he is not surprised that the decision is not being appealed. Even so, the battle is hardly over. Because the ruling only applies to the district under the presiding judge's jurisdiction—which includes only portions of California—it does not directly affect national policy-making standards. At least not yet.

William Carson, a legal scholar at the Kellogg Institute, believes that the administration may attempt to string together several such precedents and thus fabricate a "legal necessity" argument that would effectively neutralize public opinion and political opposition. "The courts can be used as a kind of trump card," he says:

> If the president can somehow showcase a few manipulated test cases—and no one in the higher courts calls his hand—he may actually be able to win this battle by default. He just needs a handful of cases to turn the tide in his favor—and despite twelve years of conservative appointments, enough liberal judges are still out there to deliver sufficient precedents.

Colonel Ronald Ray, a Marine Reserve Judge Advocate, has pointed out that such limited court precedents are often maneuvered onto the books as "sweetheart suits." These were quite common, he says, during the Carter years when litigators involved in public-interest suits were "actually in sympathy with the plaintiffs." As a result, the government's attorneys deliberately lost their cases as a means of by-passing the more laborious political process. "They were using the courtroom to legislate minority positions that they couldn't win in the legislature."

G. K. Chesterton, the great English journalist during the first half of this century, used to quip that, "There are very few things in life that a little politics can't make worse." It seems that is more true today than ever before.

Mark Jefferson, a highly decorated Vietnam veteran, "became aware of several" homosexual incidents during his twenty years of service in the Air Force. "I am strongly opposed to lifting the ban," he says. "It is not because I think the present system is perfect—not at all. I think much can be done to improve the fairness and consistency of enforcement."

He argues that politicizing the military—whether in campaigns, cabinets, or courts—does little to either solve the grave moral crisis of our times or meet the strategic needs of our world:

> It is more than just a little imprudent to usher in sweeping changes in the nature and the character of the military community merely to satisfy the political whims of the moment. You don't have to go very far or look very hard to find something in America today that politics has irreparably harmed. Surely we're not foolish enough to inflict that kind of dystopic social engineering on the armed forced too.

Surely not.

5

THE SEVEN MYTHS OF GAY PRIDE

MARLIN MADDOUX

With all the rhetoric clouding the issue of homosexuality in our society, it is very difficult to ever get to the reality. But conservative media pioneer Marlin Maddoux has made a career out of moving past the facades and revealing the heart of the important issues of our times. In a remarkably succinct fashion, this veteran author, broadcaster, and businessman brings his fresh analysis to the current crisis.

Facts are the missing ingredient in the debate about homosexuality raging in every part of our society today. And obscuring the facts is a key tactic in the homosexuals' self-described plan to "overhaul straight America."

So here they are. Facts. Truth has the same importance in the effort to preserve personal and social welfare that oxygen has in a person's effort to breath. Let's not allow the homosexuals to cut off our nation's supply of life-giving truth. With that in mind, here is the reality behind the seven deadliest myths being pushed today:

MYTH 1: Homosexuals Compose 10 Percent of the Population

How many times have you heard this one? The problem: the figure is off by a factor of anywhere between *five-hundred to one-thousand percent!*

The "10 percent myth" stems from the badly flawed sex studies by Alfred Kinsey in 1948 and 1952. While portraying his survey as "representative" of the population, Kinsey is now known to have tilted the numbers by using a high percentage of responses from male prostitutes, prison inmates, and sex deviants. Recent data, from better surveys, paints a very different picture.

For example, in 1990 the National Opinion Research Center of the University of Chicago released an extensive study which estimated that active homosexuals and bisexuals were, *at most,* 1.5 percent of the population. This confirmed results published in 1989 by the *British Medical Journal,* as well as a U.S. Census Bureau study published in the March, 1989 scientific journal *Advance Data.* All studies took into account any factors that would cause under-reporting of homosexual conduct. These are the facts. Christians need to get them before the public eye.

MYTH 2: Homosexuals Are "Born That Way"

The truth is that none of the biological and statistical studies suggesting a genetic cause for gayness have been replicable and most have indulged in faulty logic.

In one study, Dr. Simon LeVay, a homosexual biologist, claimed to have found a tiny difference in the brains of homosexual and heterosexual cadavers. Further investigation, however, revealed that the sample size, only 43 men, was far too small to be reliable. Also, the men whose brains were sampled had all died of AIDS, which could have altered brain structure. Finally, LeVay couldn't verify that the "heterosexual" samples were in fact heterosexuals—the fact that they had died of AIDS actually suggested they were homosexual. In other words, Dr. LeVay was probably comparing homosexual with homosexual brains! This, of course, would prove nothing about the differences between gay and straight men.

In another study, it was found that if one set of twins was homosexual, his or her twin had a better chance of being homosexual as well. A biological link? That's what the media concluded. Yet the same finding would also be true if the causes were related to family influences, not a more iron-clad genetic blueprint. The truth is that extensive psychological research and counseling have already uncovered the most probable causes of homosexuality, none of which are genetic.

MYTH 3: Homosexuals Can't Change

Tell that to former homosexual John Paulk. Once a female impersonator, Paulk is now a happily married heterosexual and a counselor for Love in Action, an organization helping men and women who wish to leave the homosexual lifestyle.

Or tell it to former homosexual John Smol, director of Love in Action, who is also now a happily married heterosexual.

Or tell it to the thousands of homosexuals who are leaving the gay and lesbian lifestyle through ministries such as Exodus International, or counseling by psychologists such as Dr. Joseph Nicolosi, Dr. Elizabeth Moberly, or Joe Dallas.

Because homosexuality is a learned behavior, it can be "un-learned." The proof is in the lives of people like these.

MYTH 4: Homosexuals Are Not Likely to Molest Children

Homosexuals often cite studies showing that most molesters are heterosexual. And that statistic is true—but it's only half the story. The relevant fact is that homosexuals molest children at a far higher per capita rate than heterosexuals. That means your children are at a much higher statistical risk to be molested when under the supervision of a homosexual.

A survey of documented studies of molestation indicated that homosexuals and bisexuals have committed more than one-third of all child molestations, despite being only 1.5 percent of the population. According to the Family Research Institute (FRI) of Washington, D.C., which conducted the study:

- Homosexuals are at least 12 times more likely to molest children than heterosexuals.
- Homosexual teachers are at least 7 times more likely to molest a pupil.
- Homosexual teachers have committed at least 25 percent of all molestations of pupils.

The FRI's pamphlet summarizing the study is recommended reading. These are crucial facts to consider the next time your school board, Scout district, or YMCA considers doing away with hiring discrimination based on sexual orientation. Your well-informed voice will need to be heard.

MYTH 5: The Gay Lifestyle Is Loving and Healthy

According to former homosexual Paulk, "The gay agenda in the media today is trying to put out a message that the majority of the gay community is a loving and caring and monogamous group of people."

The reality is starkly at odds with the image.

"The majority of homosexuality, especially male homosexuality, is created around anonymous sexual encounters," he testifies in the 1992 film, *The Gay Agenda.* "Long-term partners are rare—*very rare.* Most people don't even know where these activities are taking place and if they knew they'd be shocked, because they're going on right underneath your nose."

Paulk lists common places for encounters as park shrubbery, public restrooms, mall department store restrooms, adult bookstores, and the group-sex environment of "bathhouses."

Do statistics back up his claim?

One widely regarded study had a representative sample of male homosexuals keep sexual diaries. On the average, these gay men had 106 partners per year! A later study by another scientist indicated an average of 47 partners per year. No study—even in the height of the AIDS panic—indicated a drop to an average of less than 10 per year. And the best estimates of average partners per *lifetime* are between three to five hundred!

What do gay men do? Studies show that all engage in oral sex. Nine of 10 engage in rectal sex of various forms which produce incredible physical damage and chronic health problems (we decline to mention the details here). Eighty percent admit to sex of various forms using solid human waste. In perhaps the largest survey of gays, 23 percent admitted to sex using urine in assorted techniques. Thirty-seven percent admit to sado-masochism—in other words, torture. Many abuse drugs.

What does this do to gay men? The results are death and disease. Homosexuals are 8 times more likely to have hepatitis than normal adults, 14 times more likely to have syphilis, and 5,000 times more likely to have AIDS. According to a massive 1991 study by the Family Research Institute, the median age of death for a homosexual male *not having AIDS* was only 42 years, with a mere 9 percent living to old age.

Of 106 lesbians studied, the median lifespan was only 45 years, with 26 percent living to old age.

The myth: loving, caring, monogamous, fulfilled.

The reality: twisted, lonely, diseased, dying.

These are the people who desperately need our help, not our approval.

MYTH 6: Homosexuals Are a Discriminated Class

The three most common forms of discrimination are income, education, and profession. Have homosexuals been widely victimized in these areas?

Actually, homosexuals are doing much better than the general population in all these areas. They are also much better off than a true discriminated minority, American blacks.

This also explains why gays can mount such a high-powered, high-dollar political and cultural campaign despite being a tiny segment of the population.

MYTH 7: The Bible Allows for Homosexuality

The simple fact is that the Bible condemns homosexuality for any reason. Passages against homosexuality abound in both the Old and New Testaments. Romans 1 portrays it as a sin which is a key aspect of a domino effect of social depravity which entails harsh judgment.

While Jesus did not directly speak against homosexuality, He *did* strongly endorse the entire Old Testament Law which made homosexuality a capital crime. And, He gave blanket pre-approval to the coming message of His disciples, who branded all homosexuality as a fearful sin.

The Biblical reason for this attention to homosexuality is that it is unnatural. It flies in the face of a primary dynamic of God's creation. Unlike celibacy, homosexuality is a total *reversal* and *rejection* of the most important of human roles. Yet gay "evangelicals" will still attempt to evade this obvious point.

Many seek to present the Biblical injunctions against homosexuality as referring only to homosexual prostitution, not monogamous, "meaningful" homosexual marriage. Yet the context of Romans 1 makes no such clarification, implying that the condemnation is universal, not limited. Also, sex of any kind is only condoned in the Bible within marriage, and marriage is *always* explicitly described as heterosexual by original design.

WE MUST ACT NOW

The stakes in this battle are extremely high. As men and women are lured into choosing the gay lifestyle, they usually drift from deceived to diseased to dead. Truth is, their only chance is to escape, survive, and find real love.

More to the point, no society has ever been known to reverse the slide into all-out depravity once homosexuality became acceptable in the minds of the general culture. No society that completes this slide has ever been exempt from quickly ensuing destruction.

President Clinton is obviously in the thrall of the gay and lesbian lobby. So is the dominant liberal media. So are many courts. So are the public schools.

Whether middle America crosses the line may be up to *our* ability to get the *facts* to the confused masses who are being hammered with the seven myths of homosexuality on a non-stop schedule.

•

6

MANUFACTURING RESPECTABILITY

GARY DEMAR

Just twenty years ago, the proposal to lift the ban on homosexuals in the military would have been laughed right out of Washington. No one would have given it a second thought. But today, a new aura of respectability surrounds the whole subject—so that anyone who dares not give it serious consideration is quickly labeled either a bigot or a sad relic of the past. In this penetrating study, popular author and speaker Gary DeMar examines the movement that made that turn of events possible.

Turning the case for homosexuality into a crusade for civil rights is a brilliant tactic. Large groups of once-disenfranchised Americans—racial minorities, women, and the physically impaired—are being called on to join what is being called the next phase in the cause for full civil rights. The homosexual lobby claims that to ignore their struggle for equal rights will stall the many advances made in behalf of minority groups that have fought long and hard to gain equality. In fact, the claim is made that the civil rights movement will be set back if opponents to homosexuals in the ilitary get their way. Homosexuals remind us that Harry ruman faced similar opposition in 1948 when he mandated desegregated army.

By adopting this tactic, the homosexual lobby has capured the high ground by formulating a reasoning process hat ignores several key differences between the two situtions. The logic is deceptively simple, as this example demnstrates: "Substitute 'gays' for 'blacks,' and you've pretty nuch got the situation facing President Clinton. The same horus of protest, resistance to change, fear of the unknown." ounds reasonable, doesn't it? But is it?

The homosexual lobby has skillfully maneuvered the deate to their advantage, constantly putting those opposed to gay rights" on the defensive. Recent history attests to their bility to force compliance in acceptance of their rhetoric.

ILENCING THE OPPOSITION

hose who speak out against special rights for homosexuals are orcefully denounced as bigots. Martin Luther King III was orced to apologize to "Gay-rights leaders" for remarks he nade while speaking in Poughkeepsie, New York, in 1990. King had stated that "something must be wrong" with homoexuals, certainly an understatement and an accepted evaluation ust a few short years ago. Soon after the homosexual "leaderhip" unleashed their wrath upon him, King retracted his renarks, calling them "uninformed and insensitive."

Andy Rooney, a syndicated columnist and commentator n CBS's *60 Minutes,* also learned the hard way what it neans to anger the homosexual machine. In a December 989 CBS Special, *The Year with Andy Rooney,* Rooney tated that "There was some recognition in 1989 of the fact hat many of the ills which kill us are self-induced. Too nuch alcohol, too much food, drugs, homosexual unions, cigarettes. They're all known to lead quite often to prematur death." The militant homosexual lobby leaped into action an Rooney was suspended for three months. His comments o *60 Minutes* and an allegedly racist remark in a "gay" maga zine precipitated the suspension. He was later reinstated afte just thirty-two days through a great deal of support from irat viewers who believed CBS had caved in to the pro-homosex ual lobby. Rooney later was pressured by the "gay lobby saying that he "'learned a lot' about how hurtful his unir formed comments on homosexuality had been."

"Gays on prime-time TV! Lesbians in *Redbook!* Comi book supermen bursting out of the closet! From the halls c Congress to the halls of ivy, gays and gay issues—once mer tioned in whispers, if at all—are being discussed, out lou and often." The homosexual lobby has targeted every institu tion with the goal to remake the gay way of life into a norma and acceptable lifestyle. Since the early 80's the homosexua lobby has pushed for a more sympathetic treatment of the hc mosexual lifestyle in television and the movies. Scripts ar often changed by writers to accommodate the sensibilities c the Los Angeles-based Gay Media Task Force. Homosexual are now depicted as caring individuals who simply happen t like people of the same sex. Anything that would show th perverse nature of the homosexual lifestyle is censored. Ho mosexuals are the most effective and well-organized of th special-interest organizations lobbying the television industry Gary Dontzig, co-executive producer of *Murphy Brown* an co-writer of an episode with a gay character, is aware that th realities of the homosexual lifestyle must be masked in orde to gain acceptance from the public:

> In terms of what's been happening in this country with homo phobia, we felt it was important to show a healthy, well-ad justed, wonderful and openly gay character. . . . When (people

see a gay-pride parade on the news, it's the guys who are overt. Why not put somebody on who the audience could see and say, "Hey, this is a person."

Television deals in myth. The television portrayal of ho-nosexuality is a deliberate attempt to mislead the public. On he other hand, a gay-pride parade is reality.

New York City now gives unmarried and homosexual ouples some of the rights enjoyed by married couples. hose living together can register as unmarried "domestic artners." Those who register will be entitled to the same npaid leave that in the past has only been available to mar-ied city workers. Housing will also be opened to "domestic artners." Again, normalcy and legitimacy are the goals.

The Boy Scouts have been pressured to allow homosex-al Scoutmasters to be troop leaders. The Boy Scouts have efused. The homosexual community, showing its own brand f bigotry and heterophobia, has responded by pressuring ong-time donors to cease supporting the organization. The nean-spirited attitude of homosexual activists goes beyond he denial of financial support. Yale University's student-run ommunity service organization has denied membership to a 3oy Scout troop and Cub Scout pack because of the national 3oy Scout organization's policy of barring homosexuals rom membership.

The goal is simple: Homosexuals have to convince the American public that homosexuality is a legitimate lifestyle hat should be tolerated in a modern, pluralistic society. In act, toleration is not the final goal. The stigma of depravity nust be removed. Changing the law regarding gays in the nilitary is one more step in the process of manufacturing espectability for the homosexual lifestyle. Saying that it's okay to sodomize a willing neighbor, however, does not nake it right. Getting your non-homosexual neighbor to agree with your lifestyle does not make it right, either. And although forcing the hand of Congress does not make homosexuality right either, it does make it legal.

THE NEW BIGOTS

A bigot is now someone who speaks out against perversion. Bigotry used to be directed against those who viewed various racial and ethnic groups as being inferior by nature—a preposterous notion. Blacks, Hispanics, Caucasians, Orientals, and Indians, for example, are what they are by birth. There is no changing one's race, and there are no ethical implications associated with any racial or national identity. A black man, for example, is no less a human being than a white man. The Bible asks it this way: "Can the Ethiopian change his skin or the leopard his spots?" (Jeremiah 13:23). The answer is no. An Ethiopian should not have to change his skin since race has nothing to do with his status as a human being created in God's image. The Bible does not divide people by race, contrary to what some people might think. Men and women, no matter what their race or nationality, are evaluated by how they act.

On the other hand, a homosexual is what he is by choice. Unlike the Ethiopian who cannot change the color of his skin, the homosexual can change his behavior, if he so chooses. The homosexual community is the only minority group seeking special rights in terms of sexual practices. They encourage one another "to cultivate and celebrate their sexuality." A homosexual's sexual *practice* is his identifying trait. Art Positive, a homosexual protest group, "stands for militant eroticism." William Dobbs, founder of Art Positive, states that his group will "put our images and our culture out there for everyone to see." Simply put, homosexuality is nothing more than a deviant "pro-sex movement."

In the Bible, the sin of homosexuality is listed along with the sins of fornication, adultery, theft, covetousness, and drunkenness—acts which can be forgiven in Christ. In order to receive forgiveness for these sins, however, God requires a change in behavior from those who used to practice such sins (1 Corinthians 6:9–10). "And such were some of you; but you were washed, but you were sanctified, but you were justified in the name of the Lord Jesus Christ, and in the Spirit of our God" (v. 11).

COMPARING APPLES AND ORANGES

Anti-discrimination laws which protect racial minorities, women, and the disabled are in a different category from laws protecting those who practice sodomy. Homosexuality is an action, an admitted lifestyle. A homosexual is defined in terms of what he or she *does.* Anti-discrimination laws are not designed to protect lifestyles. Human rights laws protect what a person is, not what a person does. The homosexual community is beginning to realize this, so they are reformulating the debate to include biological or genetic data to support their contention that their behavior is equal in kind to heterosexual behavior. But this will prove very little. The act of sodomy is perverse whether or not it is genetically determined. Finding a genetic cause for homosexuality does not mean that the act is normal or moral. Scientists are trying to link behavior to genetic causes—from crime to addiction to alcohol. Even if links are found, this will not mean that they are fit to serve in the military. Millions of dollars are being spent to remedy genetic flaws that cause debilitating illnesses.

One must consider the objectivity of someone who claims that sticking a penis in a sewer is akin to a man and woman having intercourse. The case for a genetic cause might have merit if, and when, conception takes place in a man's rectum.

Consider the behavior of Tommy, "a tall, well-mannered boy of fifteen" who is undergoing therapy for numerous "sex crimes." He has admitted having frequent intercourse with his nine-year-old step-brother over a three-month period. "The behavior probably would have continued, Tommy said, except his stepbrother was caught having sex with an even younger child and told authorities he learned it all from Tommy." Notice that the behavior is said to have been "learned." From whom did Tommy learn it? Tommy was "sexually molested as a youngster." At first Tommy denied the activity. Then he added, "But it was sort of a relief that it wasn't a secret anymore. I wanted to stop. I enjoyed it, but I felt it wasn't right in a way."

Ultimately, homosexuality is a learned behavior that is enjoyed. Sex offenders like Tommy are required by court officials to participate in a program "that uses unorthodox therapy techniques to try to change their behavior by altering the way they think." Because of his actions, Tommy is listed as an "offender."

BREAKING THE BACK OF "INTOLERANCE" AMONG THE YOUNG

Comparing homosexuality with race, gender, as well as physical and mental impairment, has become a culture-wide strategy of the homosexual lobby. Not even children are spared these tactics. Consider how homosexual propaganda is set forth in the seemingly innocuous world of comic books and the way the press handles the topic. In the March 1992 issue of Marvel's *Alpha Flight* comic book series, Northstar, a (fictional) former Canadian Olympic athlete, decides to come out of the closet after seeing the ravaging effects that

AIDS has had on an abandoned baby. He decides to adopt the infant AIDS victim.

The editors at *The New York Times* celebrated this favorable treatment of homosexuality: "The new story lines suggest that gay Americans are gradually being accepted in mainstream popular culture. . . . Mainstream culture will one day make its peace with gay Americans. When that time comes, Northstar's revelation will be seen for what it is: a welcome indicator of social change." *The News York Times,* in order to justify its support of homosexuality, compares discrimination of homosexuals with the discrimination of blacks, women, and the handicapped.

> Marvel, beginning in the early 1960's, was the pioneer in comic book diversity. Marvel published "Daredevil," a dynamic crime fighter who was also blind. Then came "The X-Men," a band of heroes led by a scientist whose mental powers more than compensated for his confinement to a wheelchair. And with "Powerman," "The Black Panther," and "Sgt. Fury," Marvel offered black heroes when blacks in the movies were playing pimps and prostitutes.

Of course, we mustn't forget *Wonder Woman,* who brought equality to women nearly fifty years ago. Now there are many women superheroes. Northstar's hero team is led by a woman.

Northstar's self-revelation was not the first appearance of a homosexual character. The second-largest comic company, DC comics, publisher of *Batman* and *Superman,* introduced a homosexual character—the Pied Piper—and AIDS-related themes in their *Flash* series. "Future issues will have the Pied Piper bring a male date to a wedding, and discuss the importance of protecting yourself from exposure to AIDS."

The goal of parading homosexual "heroes" is to get young people—who will one day be decision makers—accustomed to seeing homosexual characters in leadership positions. Gary Stewart, president of Marvel entertainment group, had this to say about the introduction of their homosexual "superhero":

> And at the time that . . . the team was created, Northstar . . . was considered to be gay by the creator. In earlier issues there were hints that he was. There was no direct admission at that time. We believe that the only message here, per se, is the fact that we do preach tolerance. Just as you have in every day society, you have gay individuals and straight individuals. We happen to have one character in the Marvel universe, which exceeds two thousand characters, that happens to be gay.

The New York Times, being a bit more honest than the people at Marvel, took an advocacy position. The editors wrote that it was "welcome news." Since the comic book audience is made up mostly of teenagers, that group "will benefit most from discussions about sexuality and disease prevention."

According to *The New York Times,* Northstar's homosexuality should be treated like race, physical handicaps, and gender differences. There is one problem with the comparison: *homosexuality is a behavioral choice.* No one chooses blindness, racial makeup, physical handicaps, or gender. And given a choice, people with physical handicaps would like debilitations reversed. My father is handicapped. He lost his right leg in Korea. Given the possibility of change, he would choose to have his leg back.

Consider a hypothetical superhero who develops a drinking problem, and while driving severely injures or even kills a child. The sight of this maimed child forces our superhero to confront his drinking problem and seek help. This is what we would expect. But not so with a homosexual character reflecting upon the plight of an AIDS baby. Comic book editors have given in to the strong homosexual lobby screaming in their faces to treat the love of sodomy as a normal, all-American lifestyle.

The homosexual community's strategy is evident: To soften the public to adopt the homosexual lifestyle as morally acceptable and constitutionally legitimate. Pushing for acceptance of "gays" in the military is simply one more step in the public relations ploy. The homosexual community will claim that homosexuals have always served in the military. Some have even served with distinction. This begs the question. If the behavior is deviant and criminal, it does not matter how well a homosexual has soldiered. My guess is that rapists, thieves, and wife beaters have also served in the military. Some may have also served with distinction.

7

THE LEGAL FEASIBILITY OF THE BAN

JOHN EIDSMOE

There are a number of crucial issues at stake in the controversy over lifting the military ban on homosexuals. Central to their resolution are some of the stickiest legal complexities of our day. Constitutional scholar and veteran author John Eidsmoe addresses those complexities with an unerring eye. According to Lt. Col. Eidsmoe, sufficient precedent already exists to sustain any legal assault on the ban. This is clearly a battle that can be won.

Those who believe homosexuals should be allowed into the armed forces present their arguments in many forms. But essentially their arguments boil down to three: First, homosexuals have a right to serve in the armed forces without discrimination. Second, the ban is unenforceable since homosexuals serve "under cover" anyway. Third, the ban deprives the armed forces of a valuable manpower resource. We need to examine each of these arguments in order.

HOMOSEXUALS HAVE A RIGHT

Homosexuals, it is argued, are human beings just like everyone else except for their sexual orientation. They should therefore have the same rights as others, including military service. The struggle for gay rights today is comparable to the struggle for civil rights for blacks in the1950s and 60s and the struggle for equal rights for women in the 1970s. And time is on their side. Eventually homosexuals will win their struggle for acceptance; the only question is when. Unless the armed forces voluntarily change their repressive and outmoded policies, the courts will force them to do so, just as the courts forced the South to integrate forty years ago.

In *My Country, My Right to Serve,* Mary Ann Humphries declares that military service is "my constitutional right." In fact, her premise, like the second half of her title, is flawed. Military service has never been regarded as a constitutional right. Serving in the armed forces is a privilege which the government may grant or withhold, and sometimes a duty which may be required, but not a right which the citizen can claim as absolute. The privilege may be terminated when the best interest of the military and the best interest of the nation so require. Many long term soldiers are discovering this today as reductions in force go into effect.

However, some constitutional protections do apply. For example, the equal protection clause of the Fourteenth Amendment has been applied, at least in principle, to the federal government in *Bolling v. Sharpe,* by virtue of the Fifth Amendment protection against being deprived of "life, liberty or property, without due process of law."

But the Fourteenth Amendment does not prohibit all forms of discrimination. It only prohibits discrimination that lacks a sound basis. For example, nothing in the Fourteenth Amendment prohibits a state law school from admitting students who make high marks on the Law School Admission Test while rejecting those who make low scores, because it is reasonable to conclude that applicants who do well on the LSAT will do better in law school, and become better lawyers, than those who do not. Similarly, nothing in the Constitution prohibits the Air Force from accepting top-notch law graduates into the Judge Advocate General corps while rejecting those who graduated at the bottom of their classes; once again, the Air Force has good reason to believe that, in general, top law students will be better JAGs.

So the Constitution does not forbid *all* discrimination, only *unjust* discrimination.

Racial discrimination, for example, is classified as unjust because no legitimate reason for such discrimination exists, and throughout American history and that of other nations, prejudicial practices have been practiced arbitrarily, unfairly, and maliciously.

It is fashionable to compare the struggle of homosexuals for acceptance and legal rights with that of blacks in the 1950s and 60s and that of women in the 1970s. But General Colin L. Powell, the first black ever to serve as Chairman of the Joint Chiefs of Staff, rejects that comparison:

> I think the issues are quite different. Forty-odd years ago we already had blacks openly in the military and had had them for 100 years. It was a question of equal opportunity once they were in the military. And we were talking about something that was a fairly benign characteristic, with respect to skin color. With respect to gays in the military, it is, for us, a far more complicated issue that goes to one of the most fundamental of all human behavioral traits—sexual identity, sexual orientation, sexual preference.

The case law concurs that discrimination based upon sexual orientation does not fit into that upper tier "strict scrutiny" category of race.

This is primarily due to the fact that homosexuality does not involve a "fundamental" constitutional right. As the Supreme Court noted in *Bowers v. Hardwick,* certain rights are considered more "fundamental" or "preferred" than others; such rights are entitled to "heightened judicial protection." Which rights are "fundamental?" Justice White, speaking for the 5-4 majority in *Bowers v. Hardwick,* offered the following explanation:

> Sodomy was a criminal offense at common law and was forbidden by the laws of the original thirteen States when they ratified the Bill of Rights. In 1868, when the Fourteenth Amendment was ratified, all but 5 of the 37 States in the Union had criminal sodomy laws. In fact, until 1961, all 50 States outlawed sodomy, and today, 24 States and the District of Columbia continue to provide criminal penalties for sodomy performed in private and between consenting adults. . . . Against this background, to claim that a right to engage in such conduct is "deeply rooted in this Nation's history and tradition" or "implicit in the concept of ordered liberty" is, at best, facetious.

Thus, since homosexuality does not involve fundamental rights or suspect classifications, a state does not need to show a compelling interest that cannot be achieved by less restrictive means in order to win in court. All the state needs to demonstrate is that its policy on homosexuality has a *rational basis.* Furthermore, the Armed Forces is in an even stronger legal position on this issue than are the state governments, because of the unique demands of military discipline. In *Goldman v. Weinberger, Greer v. Spock,* and many other cases, the Supreme Court has recognized that the armed forces have an interest in maintaining good order and discipline; and when the military insists that a certain policy is necessary to maintain order and discipline, the Court generally defers to that military determination even where fundamental rights are involved—and as noted earlier, no such fundamental rights are involved here.

Despite all of the media publicity surrounding court cases, the plain fact remains that every court of final jurisdiction that has ever ruled on the armed forces' exclusion of homosexuals, has upheld the basic policy as constitutional.

Several courts have struck down certain aspects of the policy or the way it was administered. For example, in *Matlovich v. Secretary of the Air Force,* the Appeals Court held that, since Air Force regulations provided that in "unusual circumstances" an airman who had engaged in homosexual behavior could be retained, T. Sgt. Matlovich was entitled to be informed specifically why these "unusual circumstances" did not apply to him. The case was subsequently settled out of court, and the regulation was subsequently modified.

An Army case involved S. Sgt. Perry Watkins, who served 14 years on active duty, acknowledged his homosexual orientation from the beginning of his career, and was allowed to reenlist on three occasions even though the Army knew of his homosexuality. Around 1982 the Army revoked his security clearance and refused to allow him to reenlist, citing his homosexuality as the sole reason. The Ninth Circuit Court of Appeals held that the Army, by repeatedly allowing S. Sgt Watkins to reenlist, was estopped from discharging S. Sgt Watkins for homosexuality, because S. Sgt. Watkins had disclosed his orientation from the beginning and had built an Army career in justifiable reliance upon the Army's acceptance of his orientation.

Both of these cases involves the procedures used in discharging homosexuals; none invalidates the policy itself. In *Dronenburg v. Zech,* the Circuit Court of Appeals held that there is no constitutional right to engage in homosexual conduct and upheld the Navy's policy on homosexuality as a rational means of achieving legitimate government interests such as discipline, good order, and morale. The court stated:

> The effects of homosexual conduct within a naval or military unit are almost certain to be harmful to morale and discipline. The Navy is not required to produce social science data or the results of controlled experiments to prove what common sense and common experience demonstrate.

On January 29, 1993, in a well-timed decision which accompanied President Clinton's announced intention to lift the gay ban, a Federal District Court Judge ruled that the Constitution forbids discharging Navy personnel solely because of homosexual orientation. But again, the case involved a unique factual situation in which Meinhold served for twelve years, during which time he repeatedly acknowledged his homosexual orientation before senior officers and others. The Navy could have discharged him at any time, but for some reason took no action until 1992 when he proclaimed his homosexual orientation on an ABC television news program. As in the *Watkins* case, the court held that Meinhold was justified in relying upon the Navy's failure to discharge him in planning his future and assuming he could complete his Navy career. This factual situation, coupled with the Navy's apparent procedural errors and failure to fully articulate its rational basis, makes this a typical case. Furthermore, it is only a Federal District Court case and unlikely to set a precedent which the Circuit Courts of Appeals or the Supreme Court will follow.

Because of these federal court precedents, the Supreme-Court's ruling upholding state anti-sodomy laws in *Bowers v. Hardwick,* and the Court's many rulings upholding the authority of the armed forces to enforce good order and discipline, it seems likely that the present Supreme Court would uphold the military policy on homosexuality as constitutionally valid, probably by a vote of 7-2.

It is, of course, possible that President Clinton may appoint more liberal justices to the Court, and these justices may vote to strike down the policy. But it is rather improbable that he will be able to alter the Court enough to change the outcome on a case of this nature during these next four years. Two of the justices most likely to leave office in the near future are the Court's most liberal—Justices Blackmun and Stevens. A third justice, Byron White, has always been a centrist and thus not a major factor in the ideological sway of the Court. Replacing them with new liberal justices would not alter the basic ideological make-up of the Court.

Whether homosexuality is a choice or a condition, the ultimate issue is whether lifting the military ban would prejudice the good order and discipline of the armed forces and hinder the armed forces in accomplishing their mission. The main purpose of the military ban is not to "punish" homosexuals but to further the military mission.

THE BAN IS UNENFORCEABLE

There is no question that some homosexuals do serve in the armed forces despite the ban, and there is no question that some of them do engage in homosexual activity.

But, the fact that homosexuality persists does not mean the ban is unenforceable or should be repealed. The same reasoning could be applied to other problems: Murder is against the law, but people still commit murders; therefore the ban on murder is unenforceable and should be repealed. Theft is against the law, but people still commit theft; therefore the ban on theft is unenforceable and stealing should be made legal. Child abuse is against the law, but people still abuse children; therefore the ban against child abuse is unenforceable and should be repealed. In a more serious vein, the same argument is made for legalization of drugs.

Obviously, our criminal laws have not eliminated all murders, thefts, or abuse of children or drugs. That does not mean these laws are failures. The real question is, how much more murder, theft, child abuse, and drug abuse would take place if these practices were legalized? Likewise the question is, how many more homosexuals would enter the military, and how many more homosexual acts would be performed, if the ban were eliminated?

The fact is, no one knows how many homosexuals are serving in the armed forces. No one knows how many homosexuals ignore the ban and enlist anyway, or how many avoid military service because they perceive the military as an unfriendly environment for homosexuals.

Likewise, no one knows the extent to which homosexual acts are committed in the military. What do homosexuals do after they enlist? Do they continue the same practices with the same frequency as in civilian life, albeit a little more secretively? Or do they enlist, conceal their orientation, and refrain from homosexual activity? Do they abstain from sex entirely, or do they attempt heterosexual liaisons? Do some genuinely attempt to convert to heterosexuality? Do some try to abstain from homosexuality but occasionally find themselves overcome by desire? Do some continue homosexual activity, but with less frequency than in civilian life? Without answers to these questions, we are unable to say whether the ban has worked or not.

Furthermore, the fact that a law or policy has not been enforced does not mean it is unenforceable. The extent to which the ban has been enforced has varied from one service to another, and also with the various commands, installations, and individual commanders. Many commanders who support the ban have chosen to enforce it only when homosexuality is brought to their attention; in most instances they have not actively sought out evidence of homosexual activity. And whichever course they follow, commanders are subject to criticism: If they vigorously enforce the ban, they are accused of "witch-hunting"; if they don't, they are accused of "hypocrisy" or "selectivity."

Justice White noted in *Bowers v. Hardwick* that law "is constantly based on notions of morality . . ." Morality and legality are not the same, but they cannot be entirely separated. Regardless of whether homosexuality is legalized, most military personnel will continue to regard homosexuality as immoral and/or distasteful and will refrain from homosexual behavior. A few will continue to engage in homosexual acts regardless of whether such acts are legal. In between are many servicemen and women whose convictions, orientations and identities are not clearly formed, in whose minds legality and morality will be somewhat merged, and who are likely to think, "Well, if Uncle Sam says it's okay to engage in homosexuality, then why not try it?" This possibility is enhanced by the fact that a large portion of the force is composed of young people in their late teens or early twenties.

According to the limited data available at present, it appears that the ban on homosexuality does deter many homosexuals from entering the armed forces and the number of homosexuals in the armed forces would likely rise substantially if the ban were lifted. Additionally, it seems that the ban deters homosexuals who are in the armed forces from engaging in homosexual activity. While the ban may not induce a complete change in sexual orientation, the deterrent effect of the ban does appear to reduce the incidence of homosexual activity considerably. And, perhaps most importantly, the ban apparently deters many who might be partially homosexual or uncertain about their sexual orientation or identity from experimenting with homosexual activity. One reason is that, with the ban in effect, such persons are less likely to be solicited by committed homosexuals.

And if, in a particular instance of a soldier whose duty performance is exemplary and whose private homosexual tendencies have not interfered with his work, his commander decides to close his eyes to the soldier's homosexuality and take no action, this flexible enforcement does not render the basic policy unworkable.

THE BAN DEPRIVES THE ARMED FORCES

If homosexuals constitute 10 percent of the general population, and if the ban prevents all of these people from entering the armed forces, then the armed forces may very well have been effectively deprived of substantial manpower. But the argument is fallacious for several reasons.

First, the oft quoted 10 percent figure is based upon the Kinsey reports, now repudiated research from nearly half a century ago. Most experts agree that there is good reason to believe Kinsey used faulty research methods which inflated his figures, and a figure between 2 and 5 percent is more realistic.

Second, the argument assumes that all of these people are excluded. To the extent that the second argument (that the ban is unenforceable) is valid, this third argument loses validity. For if homosexuals are already serving in the armed forces and the only question is whether they will serve legally or illegally, then the armed forces are not losing any manpower resources at all.

As noted above, however, the ban does indeed deter some homosexuals from entering the armed forces; so the armed forces do lose some manpower because of the ban.

On the other hand, the number of homosexuals the military loses as a result of the ban, might be offset by the number of heterosexuals who might decide against military service if the ban were lifted. Many might object to sleeping, showering, and serving in the trenches with homosexuals, regardless of whether there is a valid basis for their objections. Many might perceive the armed forces as an immoral institution. Many parents who hold traditional values might be reluctant to recommend military service to their sons and daughters—and while one is legally an adult at age eighteen, the fact is that many young people continue to depend upon their parents for advice and counsel long after that age.

The argument also assumes that those homosexuals who are serving, or who would serve if the ban were lifted, would be good soldiers. Some would be. There is no doubt that many homosexuals have been outstanding soldiers in every respect except for their sexual orientation. But is this true of all homosexuals? Or most? Or a disproportionate number? Or do a disproportionate number of homosexuals cause problems for the armed forces?

Several recent studies indicate that about 76 percent or 77 percent of homosexuals who serve in the armed forces receive honorable discharges. In some instances they were discharged as homosexuals with honorable discharges, but in most instances they served their tours of duty and were discharged honorably with their sexual orientation never discovered. This demonstrates that it is possible for homosexuals to serve honorably. But about 98 percent of military personnel as a whole complete their enlistments and receive honorable discharges, so these figures indicate the percentage of homosexuals who have problems in the military is several times higher than average.

According to a General Accounting Office study, 24 percent of those discharged for homosexuality were in a non-occupational category (which includes patients, prisoners, and students) while only 9 percent of the military as a whole are in non-occupational categories. There could be many explanations for this phenomenon, but it might indicate that homosexuals are more likely to be in nonproductive categories.

While it is true that many homosexuals have served with honor and some have even been decorated for valor, the same could be said for alcoholics, drug addicts, and pedophiles. This does not mean such persons belong in the armed forces.

Closely related to the manpower issue is the cost of replacing homosexuals who are discharged. The GAO noted that the cost of recruiting and providing initial training for an enlisted person is estimated to be $28,226, while the cost for an officer is $120,772. The GAO therefore concluded that each time a homosexual is removed from the service, the armed forces incur at least this cost in replacing that person, not counting the cost of the investigation and discharge proceedings themselves. The Department of Defense, in its response to the GAO study, commented:

> Nonconcur. Each year the Department of Defense separates about 300,000 Service members, approximately 100,000 of whom are separated for force management reasons. Homosexuals make up less than one-third of 1 percent of that total. In estimating the cost, the GAO apparently assumed that none of those separated for homosexuality would be lost through normal attrition or for force management reasons. There also was no recognition that approximately one-half the enlisted force does not serve beyond the initial enlistment. The GAO cost estimate is, therefore, well in excess of what reasonably could be projected under normal circumstances.

Furthermore, since the total force is currently being reduced in number, if homosexuals were not discharged others would be discharged in their place. The cost of discharging them, therefore, appears to be reduced to zero.

Undoubtedly, the armed forces will lose some good performers if the ban is retained. Undoubtedly, the cost of replacing these people will be substantial, though not as high as GAO estimated. However, these costs might be more than offset by the other increased costs that could result from legalizing homosexuality. One of these might be increased cost of health care through an AIDS epidemic and other diseases, as the next chapter will demonstrate.

One related manpower factor is the possibility that the performance of homosexuals in the military would improve if the ban were lifted, because they would no longer be under the stress that results from their legal predicament and the secrecy it requires. However, if homosexuals cannot function under this kind of stress, how can they be expected to handle the stress of combat? Certainly homosexuals in the military do experience considerable stress because of their orientation and its stigma. However, legalizing homosexuality would not entirely remove that stigma. And to whatever extent it would remove the stigma, the effect its removal would have upon duty performance is purely speculative.

Finally, opponents of the gay ban resort to coercion. The armed forces will lose manpower, they maintain, because some universities will not allow them to recruit on campus so long as they discriminate against homosexuals. At present there are no statistics as to how many, if any, universities impose such a prohibition, and what effect such a prohibition has on military recruitment. It is just as possible that other schools, notably religious schools, might bar military recruiters if the armed forces hired homosexuals. In any event, military policy should be determined according to the good order and discipline of the armed forces, and military leaders should not allow themselves to be intimidated by those who would, in effect, try to deny their students the right to receive information about a military career.

And if gay rights advocates do resort to such intimidation and coercion, military recruiters have a weapon of their own—a little-known Department of Defense regulation which authorizes the DOD to cut off DOD funds to universities or university programs that refuse to allow military recruiters. There is no reason this regulation could not be used, if necessary, to protect the right of military recruiters to give, and the right of students to receive, information about a military career. It is ironic that some of those who speak so loudly about nondiscrimination and freedom of expression, will themselves be the first to discriminate and deny that freedom to others, when they may advance their own interests by doing so.

The words of David Horowitz are worth repeating, even though he is speaking about the radical fringe of the gay rights movement that may not represent homosexuals as a whole:

> It is this very principle of tolerance that homosexual revolutionaries and radicals most reject. For it is this rejection that defines them as radicals. For them, tolerance itself is repressive because it denies their most cherished illusion: that they are the authentic voice of humanity, and theirs the universal political solution.

8

JUST THE FACTS

PAUL CAMERON

For years the Family Research Institute has been gathering data and providing primary research on some of the most explosive social dilemmas of our time. In this new study, Dr. Paul Cameron, the founder of the Institute, reveals the preliminary results of a comprehensive analysis of the import and impact of homosexuals in the military. Here is where the buck stops—no hype, no exaggeration, no fear-mongering; just the facts.

As the debate over President Clinton's stated intention to admit open homosexuals into the military has taken shape, the major polling organizations have reported that sizable majorities of those currently serving, and of those who had served in the past, were opposed to homosexuals serving in the military. However, questions remain concerning how much homosexual activity occurs in the military and how much disruption such activity causes.

Some claim that no significant problems associated with homosexuality in the military exist. Nationally syndicated columnist Ellen Goodman asserted that "between 5 percent and 10 percent of the military is estimated to be gay right now" so "if showers are such a charged venue, and barracks such a threatening situation, how come the problem hasn't already wrecked morale and created dissention in the ranks? How come it's come up so rarely?"

Are problems associated with homosexual activity in the services as infrequent as Goodman believes?

Beyond opinion—which homosexual spokesmen claim is based upon prejudice—what differences have the presences of homosexuals in the military made? While on a tour of duty, how many members of the armed forces have been sexually approached by homosexuals and how many have experienced some sort of disruption because of homosexual activity?

The Humphrey-Studds study of homosexuals in the military—which recommends dropping the ban—reports that large fractions of both male and female homosexuals lied to get into the armed forces, engaged in homosexual activity while in, and frequently fraternized sexually with officers and enlisted personnel. How typical are these kinds of destablizing activities by homosexual personnel?

A 1992 survey conducted by The Retired Officers' Association (TROD) found substantial correlation between strongly negative reactions to homosexuals in the military and experiences with them that were reported as highly disruptive. Those who favored the admission of homosexuals generally reported fewer and less disruptive experiences with homosexuals. The TROD survey failed to detail the kinds and outcomes of experiences officers had with homosexuality. Were those with positive and negative opinions experiencing much the same events, while interpreting them through their pre-existing opinions? Or were their opinions shaped by the kinds of experiences they had with homosexuals in the military?

Early on, the White House charged that conservative, evangelical Christians were the major force driving the opposition to homosexuals in the military. The president's spokesman claimed that non-evangelicals did not feel as strongly or as negatively about the issue. Do evangelical Christians who served in the military have a different opinion of homosexuality and report different experiences with homosexuals?

Women are disproportionately prosecuted for lesbianism in the armed forces. Is this reflective of greater proportions of lesbians than gays in the military?

Generally, past performance in school is the best predictor of success in college. Likewise, performance on past jobs is often the best predictor of success on a new job. Since no one can foretell the future, we considered it likely that past interactions with homosexuals in the military would provide useful information about how the presence of open homosexuals would affect the military. To address this issue, we performed a systematic national random phone questionnaire survey, two random mail questionnaire surveys and a hand-delivered, anonymous questionnaire survey of those serving at accessible military bases.

METHOD

Phone Survey. We generated a systematic random phone sample of 654 adults from January 31 through February 21, 1993 from the white pages of phone books in the Washington, D.C. area of northern Virginia; the southern Maryland portion of the Washington, D.C. area; central Virginia; Cape Cod, MA; the northern part of the San Fernando Valley, CA; Napa, CA; Vallejo, CA; Redding, CA; San Francisco, CA; Seattle, WA; Boise, ID; Lancaster, PA; and Denver, CO.

Because we did not use lists of those known to have served or who were currently serving, each completed interview took, on average, about fifteen calls. About half of the potentially eligible respondents refused us an interview. Female interviewers and even male interviewers were often "shielded" by male respondents from the "explicit details" of experiences with homosexuals. Because a number of those interviewed refused to elaborate further and/or alluded to tragic events, the numbers of beatings, seductions, attempted rapes, molestations, and killings of homosexuals may have been under-reported.

Mail Surveys. 5,000 questionnaires were sent to vendor-generated "random samples" from a 530,000 member list of veterans and 5,000 questionnaires were sent to a list of 120,000 "activist evangelicals." Because almost 1,400 of the evangelical list were returned undelivered, we supplemented it with 1,000 additional questionnaires sent to the memberships of a number of conservative Southern Baptist churches. Since we were only interested in the opinions and experiences of those who had served in the armed forces, we anticipated that only about one-third of those called would qualify. The questionnaire was mailed on January 22 and responses were tabulated for this preliminary report on February 27. Of the approximately 4,500 questionnaires mailed to—and apparently received by—veterans, 460 were returned, 451 of which provided enough information to be useful—about a 10% return. Of the approximately 4,300 questionnaires mailed to—and apparently received by—evangelical Christians, 221 were returned, 214 of which provided enough information to be useful—about a 5% return. Adjusting for the expectation that the evangelical list would only have about one-third of the number of those qualified to respond as the military list, both lists generated approximately the same level of response. Respondents provided their own postage, were asked to detail their experiences and to sign their name and provide their phone number for possible follow-up to reduce frivolous responses.

Active Duty Survey. Selected active-duty individuals in Michigan, Washington, Virginia, and Colorado hand-carried questionnaires onto military posts in those states. Generally, everyone in the units in which these individuals served completed the questionnaire. No names or identifiers were employed. A total of 79 men and 4 women were surveyed.

RESULTS

Retired Military, Mail Responses. Here are the responses of the 386 men who returned partially or completely filled-out questionnaires:

Opinion: 21 (5%) strongly favored, 10 (2%) favored, 7 (2%) were unsure, 15 (3%) opposed, and 398 (89%) strongly opposed admitting homosexuals into the U. S. military.

Personally approached for homosexual relations: 11% of those who responded reported that they had been approached for homosexual relations while on a tour of duty. Of those who provided details of the approach, 3 passes or approaches were by civilians, 7 were by officers toward enlisted personnel and 4 were by enlisted personnel toward other enlisteds. All of these experiences were rated "not at all disruptive" of the mission or unit.

Experienced a homosexual incident while on a tour of duty: 87 (25%) of 350 who responded reported such an incident. Most (42 of 72) of these incidents were considered "major" disruptions of the command or unit. 28 of the incidents were initiated by or involved enlisteds, 23 were initiated by or involved officers and, with one exception, featured officers attempting homosexual relations with enlisteds.

Outcomes: 15 of the approaches, rapes, or molestations by homosexuals resulted in violence against the offending homosexual. In less than half the incidents (40) the individual committing or seeking homosexual relations was discharged. Shunning and reduction of morale were also mentioned by respondents. Nothing was done in 10 of these incidents. One homosexual enlisted man claimed that his officers and units knew of his homosexual activities and were totally accepting of them. He even sent us a copy of his discharge papers as proof.

Retired Military, Evangelical Christians. Reports of the 214 men who returned partially or completely filled-out questionnaires are summarized below:

Opinion: 4 (2%) strongly favored, 1 (less than 1%) favored, 3 (1%) were unsure, 7 (3%) were opposed and 199 (93%) were strongly opposed to admitting homosexuals into the U. S. military.

Personally approached for homosexual relations: 27 (14%) of 199 reported having been approached for homosexual relations while on a tour of duty. Of those who provided details, 8 solicitations or molestations were by enlisteds upon enlisteds, 6 by civilians, and 1 by an officer on an enlisted. More of these incidents were rated "minor" or "not at all" severe than were rated "major" disruptions.

Experienced a homosexual incident while on a tour of duty: 32 (16%) of 194 reported experiencing a homosexual incident. About half (16 of 34) of these incidents were judged "major" disruptions of the command or unit. 24 of the incidents involved enlisteds engaging in or seeking homosexual sex with other enlisteds, and 6 involved officers seeking or engaging in homosexual sex with enlisted personnel.

Outcomes: 6 of the approaches, rapes, or molestations by homosexuals resulted in the homosexual being beaten or physically assaulted for his efforts. In a little less than half of the incidents the individual attempting homosexual relations was discharged, but shunnings and reduction in morale were noted as well.

Phone Survey. The responses from the 632 men who were interviewed are summarized as follows. 3% strongly favored, 15% favored, 14% were uncertain, 18% opposed and 50% strongly opposed permitting homosexuals into the U. S. military.

Homosexual approach? 15% said that they had been approached for homosexual relations. Of those who provided details, 5 were by enlisteds on enlisteds, 1 by an officer on enlisted. 10 (23%) of these events were considered major disruptions.

Experienced a homosexual incident? 18% said they had experienced a homosexual incident. Of these, 32 (46%) were considered major disruptions. 20 were propositions or attempts by enlisteds upon enlisteds, 7 were advances by officers against enlisteds.

Outcomes: 17 of the offending homosexuals were discharged, 8 beaten or hit, 5 were shunned, 5 were transferred, and 3 were killed.

Active Duty. 3% of the 79 males on active duty favored or strongly favored homosexuals being admitted, 3% were uncertain, 18% were opposed, and 78% were strongly opposed.

Homosexual approach? 9 (12%) of 78 reported having been homosexually approached.

Experienced a homosexual incident? 28 (35%) of 79 reported having encountered a homosexual incident. 17 of these were considered major and 7 minor. 8 of these incidents involved enlisted on enlisteds. Officers attempting to sodomize, or actually sodomizing enlisteds, was reported twice. 7 of these events happened in the showers or while someone was sleeping or drunk.

Outcomes: 16 of the offending homosexuals were discharged, 2 were beaten, 1 was pushed in front of a car, but not killed.

Women. In the phone survey, 3 (14%) were strongly in favor, 6 (27%) were in favor, 3 (14%) were unsure, 4 (18%) were opposed and 6 (27%) were strongly opposed to homosexuals in the military. 2 (10%) of 21 reported having been homosexually approached and 7 (39%) of 18 said that they had encountered a homosexual incident while on a tour of duty. 1 of these incidents was considered major and 3 were considered minor. One of these female respondents indicated that she was a lesbian.

These are preliminary results, and we are continuing to poll. However, there is a good degree of agreement between all of the studies.

IS HOMOSEXUALITY A PROBLEM IN TODAY'S U. S. MILITARY?

To those like Ellen Goodman who argue that "homosexuality doesn't really affect the functioning of the military today, and it certainly won't when accepted," it is noteworthy that between 11% and 15% of men reported homosexual approaches and 16% to 25% encountered a homosexual incident during their service. About half of these incidents were of considerable significance for a fighting force, with many involving serious breaches of morale, military authority and chain of command. These same data call in question the claims of homosexual spokespersons that opposition to their open inclusion is merely based upon bigotry.

ARE HOMOSEXUALS ENGAGING IN AS MUCH DISRUPTION AND SEX IN THE MILITARY AS THE HUMPHREY-STUDDS STUDY SUGGESTS?

The Humphrey-Studds study presents a chilling view of the influence of homosexuals on the military. The interviews detailed in their study makes one wonder whether they have perhaps magnified their influence and sexual exploits. The picture from the other side of the coin—looking at them from the perspective of the non-homosexuals who serve with them—looks the same as they have presented it. The same degree of rebellion and disruption that Humphrey-Studds depicts fills the reports we received.

ARE OPINIONS ABOUT HOMOSEXUALS IN THE MILITARY DRIVEN BY BIAS OR EXPERIENCE?

Examination of the data suggests that those who have the best experiences with homosexuals—where the intrusion is not too sharp, offensive, or serious—are the most inclined toward admitting homosexuals to the military. Apparently the more frequent sour experiences with homosexuals drives the opinion of those who strongly oppose their entry into the services. In the TROD survey, those who favored or strongly favored admission of homosexuals reported only 1 major incident, 24 minor incidents, and 18 incidents that were not disruptive. Those who opposed or strongly opposed reported 180 major incidents, 98 minor incidents, and 30 incidents that were not disruptive. Thus, as in our studies, those who opposed homosexuals in the military were more apt to report negative experiences with them.

ARE EVANGELICAL CHRISTIANS THE MAJOR FORCE DRIVING THE OPPOSITION TO HOMOSEXUALS IN THE U. S. MILITARY?

The evidence we bring to bear on this issue comes from the two mail surveys—of retired military and evangelical Christian retired military. The opinions and experiences of evangelicals who served versus others who served seems far more similar than different. Recalling that the respondents who send in questionnaires probably care more deeply about an issue than those who are contacted by phone, there was only a slight tendency for evangelicals to more frequently express opinions against dropping the ban. But otherwise, their reports are indistinguishable from non-evangelicals who served.

IS LESBIANISM A GREATER PROBLEM AMONG THE FEMALES IN THE SERVICES THAN HOMOSEXUALITY IS AMONG MEN?

Our results are too limited to put great confidence in them on this point. However, a persistent, albeit uncertain effect which turned up in the small proportions of females in our surveys—women accounted for 3.3% of the respondents of our phone survey, 1.5% of the mail samples, and 4.8% of the active-duty survey—suggests that homosexuality may be more frequent among female members in service. Not only did a significant number of the incidents that men reported involve women, but women tended to be less opposed than men to homosexuals in the military. It will take larger samples of women to solidify confidence in this possible finding.

9

WHERE WILL IT ALL END?

DON FEDER

There is a domino effect at work in all social relations. For that reason, the sweeping moral and ethical changes the Clinton administration wants to effect by executive fiat in the military have implications for all of American culture and life. Don Feder is a twice-weekly columnist and editorial writer for the Boston Herald *in addition to being a nationally syndicated columnist. He believes that we may very well be venturing onto a very slippery slope—and there is no bottom in sight. How will we maintain our footing? How will we keep our equilibrium? Where will it all end?*

In the course of cultural warfare, sometimes it's necessary to step back from the fray and try to determine exactly what is at stake—repealing the ban on homosexuals in the military, a national gay rights law, local ordinances prohibiting discrimination based on sexual preference, domestic partnership laws, AIDS education in the schools.

Where will it end? Why does it matter? What will be the consequences of losing this crucial debate? Gay rights is the beginning, not the end of the movement's agenda. Ending discrimination against so-called sexual minorities is only phase one. Ultimately, the Judeo-Christian ethic must be destroyed and the traditional family delegitimized in order for the homoerotic culture to achieve its ultimate goals. Among themselves, homosexual activists are quite open about long-range objectives. A strategy statement published in the *New York Native* on January 3, 1983, is a testimony to this:

> We are essentially a radical movement, and insofar as we are successful, we do indeed need to break down dominant authority of certain traditional values about sexual relationships. Often this perception is argued in terms of a need to defend our own sexual minorities, whether they be man-boy lovers, transvestites or sadomasochists.

This point is reiterated in an article in the *Village Voice* on January 23, 1984:

> In the end, the gay alternative means a departure not just from heterosexuality, but from social orthodoxy. In other words, gay liberation is a social event. In its most moderate politics—the enactment of civil rights legislation—it has radical value because civil rights legislation means the way to acceptance, and acceptance means the dissolution of the norm.

These writers—and they are certainly well-qualified to address the matter—are in effect saying that the homosexual movement is an assault on standards. It is an effort to divorce sexuality from morality, and to obliterate the very idea that there is good and evil, right and wrong, beneficial and detrimental in the realm of sexual relationships.

Once homosexuality is accepted, it will be logically impossible to oppose any other perversion or sexual indulgence. Think of the consequences for family cohesion and social relationships. While conservatives are busy raising practical objections to specific items in the movement's agenda—for instance, we don't want homosexuals to serve in the military because it will hamper the combat effectiveness of our armed forces—the underlying premises of the movement largely go unchallenged. In other words, we are losing the philosophical debate by default. And, as goes the intellectual argument, so goes the war.

We must awaken the public to the dangers inherent in acceptance of the homoerotic world-view. For example, the toxic notion that minority status should be conferred on the basis of sexual behavior.

Regarding the debate over gays in the military, General Colin Powell—God bless him—alluded to the core issue here when he forcefully rejected the comparison of lifting the gay ban to the integration of blacks in the military following the Second World War. The Chairman of the Joint Chiefs of Staff noted that "skin color is a benign characteristic." It says absolutely nothing about disposition or conduct. Homosexuality says everything about what's in a person's heart and head. The very definition is behavior-based, i.e., a homosexual is one who performs homosexual acts.

There is an even more lethal assumption that's rarely disputed, the idea that certain behavior is innate. The homosexual movement now insists that their preference is genetic. As activists are forever telling us, they no more chose their sexual orientation than their hair color or their racial identity. If we grant this premise, if we even tacitly accept it, the debate is over.

The American people abhor discrimination in any form. If homosexuals aren't responsible for who they are—and, by extension, for what they do—then discrimination in any shape or form ultimately will be abolished. I submit that the acceptance of this doctrine is social thalidomide. It will lead directly to a morally mutant society without self-control, where any indulgence is justified and individual responsibility is considered a quaint, archaic notion. What's to stop the pedophile from insisting, "I can't help myself. This is my nature. The fact that a man wants to sleep with a little boy says nothing about the type of person he is. Stop persecuting me, accept me and demonstrate your compassion by protecting my civil liberties."

If this sounds a bit far-fetched, consider statements by Dr. John Money, Director of Psycho-hormonal Research at Johns Hopkins University, and an authority who was among the first to urge the medical profession to reclassify homosexuality as an orientation. Dr. Money writes in the *New Statesman* of August 21, 1992:

> Pedophilia is not a voluntary choice, and there is no known treatment. Punishment is useless. One must accept that pedophilia exists in nature's overall scheme, and with enlightenment, formulate a policy of what to do.

Doubtless this enlightened policy will include pedophile pride days and legislation banning bias in housing and day-care employment. Forget perversions for a moment. What about the promiscuous *heterosexual* who takes his cue from the gay rights crusade? It's not that he wants to be a sexual predator, you understand, it's part of his genetic makeup. When he sees a desirable woman he reacts like Pavlov's dog. He can no more control himself than a man in the desert who spots an oasis. Should we understand and accept him, too?

Perhaps some people are biologically programmed to be violent. Soon the man who uses his girlfriend for a punching bag will plead, "It's just my nature. When I get mad I can't control myself. Understand me. Accept me." If you think we have a problem with domestic violence now, just wait until this mindset takes hold. Tell people that they're not responsible for their actions, that their DNA made them do it, and that's exactly what you'll get—an irresponsible citizenry, and a society where people don't try to exercise a minimal degree of self-control.

A society that works requires discipline. My apologies to Woody Allen, but erotic urges must be sublimated. It also needs families, not random conglomerations of individuals who happen to be sharing living space, but men and women and their children, united by bonds of affection, blood, and convention. None of this will be possible if the homosexual agenda triumphs. Think well about what's at stake here. This is the most crucial issue which confronts us. If this struggle is lost, none of the others will be worth winning.

10

THE BOTTOM LINE

HOWARD PHILLIPS

Howard Phillips pulls no punches. He never has—from his early days in the White House during the Nixon administration to his front-line battles with both the compromising "conservatives" and the compromising "liberals" during the last presidential campaign as the head of the U. S. Taxpayer's ticket. In this trenchant essay, he demonstrates why he is one of the most consistent voices of uncompromising truth, justice, liberty, and authentic statesmanship in Washington today.

For many reasons, America's future prospects may be discerned from the outcome of the ongoing battle regarding the proposed homosexualization of the armed forces of the United States.

The military is the last major American institution to have withstood, to some degree at least, the counter-cultural assault which has captured our nation's teaching media: primary, middle, and secondary schools, colleges and universities, theater, radio, and television networks.

Only the military continues to manifest and instruct in its codes of conduct the traditional values and the Christian precepts which undergird our civilization.

This is not to say that the military has successfully withstood all assaults. Bill Clinton has now ruled, without any effective Congressional opposition, that the murder of unborn children shall occur at U. S. military hospitals. And, long before Clinton, since World War II, it has been an approved practice to distribute pornography at military facilities.

Many of the military's present problems are the result of a policy of feminization launched by Eleanor Roosevelt (with the cooperation of General George Marshall) and carried forward ever more aggressively by every succeeding president, including Ronald Reagan, George Bush, and Bill Clinton.

But the impending homosexualization of the armed forces is the greatest challenge ever posed, not just to the effectiveness, but the moral legitimacy of our nation's military.

How can God bestow His favor on an army or a nation which condones that which He declares to be an abomination?

How can Bible-believing young men voluntarily choose to participate in an institution which declares degeneracy and perversion to be socially acceptable?

How can any orthodox Christian clergyman or Jewish Rabbi agree to provide chaplain services for an institution which would require its members, not merely to tolerate, but also to approve of conduct which is an affront to God and to each of those who would be His servants?

Of course, the practical problems, as well as the moral ones, are enormous. It would only be a matter of time before exclusion of HIV-carriers would be regarded as discriminatory. Under the Americans with Disabilities Act, such is now the case for private businesses.

Military service would become particularly appealing to AIDS-infected homosexuals who, though possibly unable to secure private insurance, could quite readily stick their medical bills to the defense budget and to the American taxpayer.

Everyone involved in the armed forces would be placed at greater risk for contracting not only AIDS, but other diseases which are the by-products of sexual degeneracy.

Battlefield transfusions could be more dangerous than enemy bullets. A good night's sleep in a barracks room permeated with sodomites would be difficult to assure. Showers and bathrooms could become combat zones, and foxholes could become death chambers, without a single enemy shot being fired.

Scarce defense dollars would increasingly be reassigned to pay the medical bills of homosexual enlistees, and taxes would likely climb even higher in the process.

Only that is inevitable which is unresisted. The tragedy in America today is that neither the Republican nor the Democratic Party is willing, as a matter of principle, to oppose homosexuality.

Yes, many Republican leaders and even some Democrats have, on pragmatic grounds, taken a stand against homosexuals in the military. But, virtually without exception, ranking Republican spokesmen have accepted the formulation of the homosexuals that the issue is one of tolerance rather than one of morality.

Senate GOP leader Bob Dole, when asked for his views on the issue, indicated that he favored some kind of compromise, and that he was uncomfortable being associated with a policy which could be perceived to be discriminatory or to reflect "intolerance."

Former Republican Defense Secretary Dick Cheney has defended his appointment of a known homosexual to a position of significant responsibility on the civilian side of the Defense Department, and has argued that it is only for pragmatic reasons "for the good of the force" that he is against putting homosexuals in combat.

Even those Republican Senators who have been unwavering in their opposition to pro-homosexual policies have, with few exceptions, been unwilling to frame the issue in Biblical terms, lest they be perceived to be "intolerant," or to have violated some false notion regarding the "separation of church and state."

If Christians and conservatives rely on the Republican Party, let alone the Democratic Party, to block the homosexualization of the armed forces, the day of reckoning may be delayed and the process slowed, but the outcome will be a victory for the homosexuals. That is because neither the elephants or the donkeys establish their policy on the basis of fixed Biblical guidance or Constitutional standard.

Rus Walton of the Plymouth Rock Foundation has pointed out that Article 1, Section 1 states "All legislative powers herein granted shall be vested in a Congress of the United States, which shall consist of a Senate and a House of Representatives." Section 8 goes on to say:

> The Congress shall have the power . . . To make Rules for the Government and Regulation of the land and naval Forces . . . To make all laws which shall be necessary and proper for carrying into Execution the foregoing Powers, and all other Powers vested by this Constitution in the Government of the United States, or in any Department or Officer thereof.

Walton asks:

> Where does the President get the authority to issue unilaterally the rules and regulations for the armed forces? As the Chief Executive is it not his responsibility to carry out—through his appointees and chain of command—those rules and regulations established by the legislative branch? Who in Congress is willing to assert the institution's Constitutional prerogatives?

But here again, even if we use the Constitution as a barrier, we will, nonetheless, eventually fail, unless we recognize that the Constitution loses meaning, except insofar as it is rooted in the foundation of the Bible and the common law standards derived from it.

The urgency of reasserting such a standard has never been greater, given the fact that, with the already announced retirement from the U. S. Supreme Court of Byron White and the possible retirement of other justices, President Clinton will soon be able to nominate justices who will likely rule that Fourteenth Amendment rights of privacy are sufficient to strike down not only state laws prohibiting sodomy, but even those provisions of the Uniform Code of Military Justice, which take a similar stand.

No member of the United States Senate, Republican or Democrat, opposed the confirmation of Janet Reno to be Attorney General of the United States—despite the fact that Reno has consistently declared her unequivocal support for abortion and is, therefore, unwilling to uphold the first duty of the law, which is to prevent the shedding of innocent blood.

If Republican Senators will not take a stand against a pro-abortionist becoming Attorney General, how can we look to them to oppose prospective Supreme Court justices who believe it is unconstitutional to treat acts of homosexual perversion as criminal? How can we rely on them to make the moral case against homosexualizing the military?

Any person who is serious about preventing the complete homosexualization of our culture, not just of our military, should ask himself or herself "What responsibility do I bear for those in whom I invest my vote? Am I not responsible for supervising my elected officials, as well as my employees and my children? Am I not complicit in advancing the homosexual agenda if I support senators, congressmen, and presidents—even those who are the lesser of two evils—who are prepared to accept the formulation that this is an issue of tolerance versus intolerance, rather than one of good versus evil?"

In 1992, leading Christian and conservative activists embraced the reelection of George Bush despite the fact that he proposed massive increases in funding for Planned Parenthood, for "safe sodomy" subsidies, for the National Endowment for the Arts, for the Legal Services Corporation, and for other government entities and activities which promote abortion and homosexuality.

It is seldom the case that our "employees" in government will adhere to a standard higher than that on which we insist, especially when there are political inducements to move in an opposite direction.

Preventing the homosexualization of the armed forces and of the nation is not so much the responsibility of your senators and congressmen as it is your responsibility. It is not too late for you to let candidates and incumbents know what you will require of them before you confer your support.

11

THE SPIRITUAL STATE OF THE UNION

D. JAMES KENNEDY

We are living at a time when there are few statesmen, few authentic intellectuals, and fewer still genuinely good men. Dr. D. James Kennedy is unique in fulfilling the demands of all three roles. As a renowned author, broadcaster, and national leader, he has brought his insight and wisdom to bear on many of the most pressing issues of our time. In this essay, first delivered to the congregation he pastors in Ft. Lauderdale, Florida, he assesses America's current spiritual status.

According to Psalm 11:3, "If the foundations be destroyed, what can the righteous do?" Many a theologian and commentator has pondered over those words. Everywhere we look today, we see the foundations of old being destroyed.

Thankfully, there are certain foundations which cannot be destroyed:

First, the foundation of God's throne standeth sure. It cannot be moved. His throne is above the mountains and the clouds.

Second, the foundation of Christ and His cross, upon which our salvation is rested, standeth sure. It cannot be shaken.

Next, the foundation of our hope, which is in the resurrection of Jesus Christ, standeth sure; it cannot be moved. All of the attacks of all the critics for two thousand years have failed to even put a dent in it.

And finally, the foundation God has given of the salvation of His own also stands sure. "Nevertheless, the foundation of God standeth sure, having this seal. The Lord knoweth them that are His" (2 Timothy 2:19). His own people, His elect, those of His choosing can never be shaken; they stand sure.

But there are, obviously, as our Scripture would tell us and as history abundantly confirms, other foundations which have been, and can be shaken. These are the temporal foundations of this life. Much of our history is the story of the shaking of those foundations.

This nation had a foundation. It was founded upon the Word of God. George Washington, the father of our country, the first president, declared that it would be impossible to govern without God and the Bible. John Adams, another of our founders, declared it would be impossible to govern without God and the Ten Commandments. So that was the foundation laid by the men who gave to us this great nation.

We have now come to a critical juncture. I am speaking not of politics, but of spiritual and moral issues. Let me point out, this is not a matter of the church obtruding itself into politics, but rather, it is a matter of politics obtruding itself into the moral and spiritual realm of the church.

I saw a cartoon recently which showed a couple of ladies walking down a busy street before Christmas time. They stopped to look in one of the windows of a department store where a creche was displayed. One of the ladies said, "Humph, now the church is trying to butt into the business of Christmas."

There are some people today who don't realize that politics and government have, for the last few decades, been doing something it never did before, and that is obtruding itself into the realm of the spiritual and the moral. Therefore, I feel obliged to speak out about it. It is not something I like to do, but someone needs to declare the truth. We need to declare the truth, when it is popular and when it is not popular. The Old Testament prophets spoke out against the evils of both Israel and Judah, regardless of how they were received by the people. So, dear friends, today I speak the truth to you from my heart. You may not like it; you may hate it. You may not like me for telling you but, nevertheless, I am certain it is true.

We now have, for the first time in the history of America, a new administration with a new platform and with new promises that it intends to codify into law a clear violation of the commandments of God. This is an effort to destroy the foundation upon which this nation was built. We see that there has been the promise to pass a national gay and lesbian rights act which would grant minority status—civil rights in minority status—to those who are practicing the homosexual lifestyle. It is not, in the opinion of many and I believe most, a minority situation, but one which is chosen as a lifestyle. That, of course, is a violation of the seventh commandment which forbids all manner of sexual immorality.

The Bible clearly and repeatedly condemns homosexuality. For example, we read, "Thou shalt not lie with mankind, as with womankind: it is abomination" (Leviticus 18:22). Or, in the first chapter of the book of Romans, in the New Testament, it says:

> Wherefore God also gave them up to uncleanness through the lusts of their own hearts to dishonor their own bodies between themselves . . . .For this cause God gave them up unto vile affections: for even their women did change the natural use into that which is against nature. And likewise also the men, leaving the natural use into that which is against nature. And likewise also the men, leaving the natural use of the woman, burned in their lust one toward another; men with men working that which is unseemly, and receiving in themselves that recompense of their error which was meet (Romans 1:24, 26, 27).

It is called abomination, uncleanness, dishonorable, vile affections, unseemly, and an error, and it is a violation of the commandment of God. For the first time, to enact into a law those things which are clearly violations of God's Ten Commandments is to attempt to destroy our nation's foundation. "It is impossible to govern without God and the Ten Commandments," said John Adams—and we are now going to attempt to do so!

Let me say this: I do not hate anybody. I do not hate people who have abortions. I do not hate people who are involved in the homosexual lifestyle. I love them and I try to help them. I do not, however, approve of their conduct. The Scripture is clear; we are to love the sinner, though we are to hate the sin. Too often, that is forgotten. God makes it very clear that He is not pleased with such lawbreakers as these.

We are told that the ban against homosexuals in the military is going to be lifted. General Colin Powell has expressed his strong disagreement with that. Each one of the heads of the various military groups in our country—the Joint Chiefs of Staff—are unanimously opposed to that and have even hinted that they will collectively leave, as will many other high ranking officers in the army if this is passed.

Someone said it would be like taking a twenty-year-old, red-blooded virile heterosexual man in college and telling him he is going to live in the women's dormitory. He is going to sleep with them, undress with them, dress with them, eat with them, shower and bathe with them, and he's going to be doing that for the next four years. If that young man has any libidinous thoughts, his idea is that he has just entered paradise. And, my friends, it would be the same way in the military. The dangers on the battlefield, foxholes, tanks, blood transfusions, the spread of AIDS . . . there are so many reasons why homosexuals in the military would be wrong.

Of course, I know there are those who say they were born this way and they cannot help it. I also know there have been a couple of supposed experiments that have seemed to support that. I have also read the scientific evidence that completely debunks those experiments and shows them to be fraudulent. Therefore, I do not believe people are born that way, and I believe it is possible that they can be delivered from that lifestyle. I read only recently that the statistic repeated time after time from the Kinsey Report of 10 percent of the population being homosexual has been proven to be completely fraudulent. Apparently, many of the people Kinsey interviewed were prisoners in our penitentiaries. Five percent of them were male prostitutes. There were people who were sex offenders. These were the kind of people that were being interviewed. It is estimated by those who do work in this area that 2 percent of the men in this country are male homosexuals.

My friends, according to surveys, 2½ percent of the men in this country declare that they formerly lived in the homosexual lifestyle and gave it up and now are practicing heterosexuals. Some say that no one can leave the homosexual lifestyle. That is a lie. There are more ex-homosexuals in America than there are practicing homosexuals. I interviewed four ex-homosexuals on my radio program one time and listened to their stories. They can be delivered.

I would like to say to any Christians who are involved in that lifestyle: You believe that God created you. Do you believe the Word of God? It clearly states that God condemns homosexuality. He says in 1 Corinthians 6:9 that neither adulterers nor fornicators, nor effeminate—the word is homosexuals—shall have any inheritance in the kingdom of God. My friend, if you claim to believe the Word of God, you cannot possibly believe that God created you that way and then is going to condemn you for the way He created you.

Many psychologists have said that homosexuality is something which is acquired in the early years of life, primarily because of the relationship with one or another of the parents. There are many who have abandoned it and God has blessed them. You can too. The last thing in the world we need is the sanction of law upon it. I pray God will deliver us from that.

If I am convinced of any one thing, after thirty-seven years as a Christian, it is this: You cannot violate the commandments of God with impunity. Well, it has been said that we cannot break the commandments of God, we can simply break ourselves upon them. And we are doing just that. My Scripture text of Psalm 11:3 was written during the time of Absalom and his rebellion—when he threw David, his father, off his throne. David's counselors were saying, "Flee as a bird to your mountain, for lo the wicked bend their bow, they make ready their arrow upon the string, that they may privily shoot at the upright in heart" (Psalm 11:1b, 2). Ah yes, they were doing just that. The foundations of government in Israel had been completely overturned. The king had been thrown off his throne, and his rebellious son, Absalom, was taking his place. "If the foundations be destroyed, what can the righteous do?" That was the plight. David answered:

"The Lord is in his holy temple, the Lord's throne is in heaven" (Psalm 11:4).

When Abraham Lincoln was slain, at least half of the country went into great mourning. There was consternation, confusion, despair, hopelessness. In New York City, a great crowd of people had gathered together and were expressing their despair and their hopelessness after what had happened to their great leader. Suddenly, a man climbed up the stairs of a building where he could look over the crowd, and he shouted with a loud voice: "The Lord reigns in Washington." The people were silenced as the meaning of that statement settled down upon them. Slowly they began to disperse and go about their business.

Yes, the Lord is in His Holy Temple; the Lord's throne is in the heavens. It cannot be shaken by any wickedness of men. But the Lord is very real and we read that "His eyes behold, His eyelids try, the children of men. The Lord trieth the righteous" (Psalm 11:4b, 5). One of the reasons we have such troubles in our world is to test the children of God, that they will be strong, that their faith will be in Him. David did not despair and we are not to despair, whatever comes upon us in this world. The Lord reigns. That is our hope. That is our trust. Blessed is the man that trusteth in the Lord and cursed is the man that trusteth in the arm of flesh. "Cursed be the man that trusteth in man," the Bible says in Jeremiah 17:5. Our trust is not to be in any president, in any Congress, in any court. Our trust is to be in the living Lord God Almighty.

> The Lord trieth the righteous: but the wicked and him that loveth violence his soul hateth. Upon the wicked He shall rain snares, fire and brimstone, and a horrible tempest: this shall be the portion of their cup (Psalm 11:5–6).

God has been doing that, and I believe God is going to do it even more unless this nation turns. We cannot violate the laws of God with impunity.

People ask the question: Is AIDS a plague from God upon homosexuals? I believe that AIDS is unquestionably from the hand of God. God ordains all things that come to pass. He either directs or permits everything that happens. He is the Sovereign Lord God Almighty. In the Old Testament we read where ten plagues fell upon the nation of Egypt. I believe that it is not merely upon homosexuals, but upon all immoral persons. There are promiscuous heterosexuals in this country who are going to die by the millions unless they repent and turn from their wicked ways. An article from *U. S. News and World Report* estimates that by the end of this decade, health care costs in America could be as high as $514 billion, just from AIDS—that it will totally destroy our country's health care system. The World Health Organization (WHO) estimates that by the year 2000, one-half of all the people in the continent of Africa could be infected with HIV. It is a pandemic, a world plague and there is nothing that will stop it . . . but morality. The answer: Abstinence before marriage and fidelity within marriage.

Someone recently asked, "Well, how could it be a plague from God, because innocent people are dying. The hemophiliacs have almost all contracted this disease." Tragically, look at the situation with Egypt, look at the situation with all of the plagues that fell upon Israel later because of their disobedience, and you will see that many millions of innocent people died. We bring down upon our own heads—because of our sins—great punishments from God, and we bring them not only upon our own heads, but upon others as well.

Does not the Bible tell us that unto the third and fourth generation we will bring evil upon the children and grandchildren and great grandchildren of our family because of our wickedness and violation of the commandments of God. Yes, I think God is not pleased with the violation of His commandments and He can make that very plain. He will rain down His torments upon people who flagrantly abuse and violate His laws. God does not do things instantly, in most cases, but He does them thoroughly. The mills of God grind exceedingly slow, but exceedingly fine, and I guarantee you that all of the promiscuity in this country will be absolutely abolished by the Almighty God, one way or another.

We have violated His commandments in stealing from our children and grandchildren through unrestrained deficits and spending, through following the Soviet Union down the socialistic, welfare path to get something for nothing—our so-called entitlements, which are nothing of the sort. We shall experience its inevitable consequence. The Grace Commission told us during Reagan's presidency, that at this rate, by the end of this decade, the interest on the federal debt will equal the entire federal budget, which means that our federal government is going to go completely bankrupt. This is going to cause absolute catastrophe financially upon people throughout this nation.

We are going to discover, one way or another, that there is no free lunch. The Soviets had to discover that the hard way. They had to wait until their entire nation was destroyed; their economy a total shambles—the average income of the Russian was $20 per month. "We have destroyed ourselves," they said. Egypt had to wait until finally the counselors of Pharaoh said that this is the finger of God; know ye not that Egypt is destroyed.

What can we do? First, we can go on being righteous. We can become more righteous. We can live godly lives and shine as bright lights in the midst of a dark world. Second, we can pray. We can pray for those who have the rule over us. We can pray that God will move in their hearts and cause them to repent and follow the laws of God. Third, we can become active in the things of this nation. Over seventy-five years ago, the running of the government was turned over to people who were not believers. Millions of Christians said they would have nothing to do with politics because it is dirty business. They didn't want to get their hands dirty. Well, my friends, that dirty business is smearing itself over the whole country, and if godly people will not reign, then ungodly people will. The Bible makes it very plain that when the wicked are in power, the people mourn. Until we learn that lesson, we are going to mourn. Lastly, we can witness for Christ. It is because we have failed to obey the Great Commission that so many people in this country today do not take God's Word seriously. They do not love God, they do not trust in Jesus Christ because we in the Church have failed to tell them the Gospel—the Good News that Christ came to save; that those who will repent and put their trust in His cross will be given the free gift of eternal life; that heaven is free to all of those who will invite Him into their lives. We have kept the Gospel to ourselves; therefore, the unrighteous and the ungodly have abounded and multiplied in this country—instead of the believers multiplying, as the Bible says we should.

If we would do these four things, I believe we could turn back the tide:

- If we could determine to live more godly lives;
- If we would give ourselves to godly prayer;
- If we would become active in the affairs of this nation;
- If we would be faithful witnesses of Jesus Christ.

As 2 Chronicles 7:14 so aptly states:

> If my people, which are called by my name, shall humble themselves, and pray, and seek my face, and turn from their wicked ways; then will I hear from heaven, and will forgive their sin, and will heal their land.

May God grant that healing to this needy land!

12

THE END OF THE HONEYMOON

Cal Thomas

Probably no new presidential administration has been more widely lauded or more wildly heralded than that of Bill Clinton. Political prognosticators marveled at his savvy, his energy, and his intellectual acumen. It was predicted that his honeymoon would last throughout his first 100 days. My, how things change. As respected syndicated columnist Cal Thomas points out, Bill Clinton is mired in difficulties that could undo his presidency from the outset—and central to his woes is his crusade to lift the ban on homosexuals in the military.

Bill Clinton's insistence on going ahead with his plan to lift the ban on homosexuals in the military is a colossal mistake that could severely weaken the recently refurbished armed forces and irreparably harm public confidence in the new President.

It might also undermine his re-election prospects.

In a memo to the president, Defense Secretary Les Aspin said that lifting the ban must simultaneously satisfy three criteria: First, it must end "discrimination" against homosexuals in the armed services, second, it must "maintain military morale and discipline," and third, it must "be defensible in Congress, so as not to be overturned."

These are mutually exclusive goals. Senate Armed Services Committee member Dan Coats has taken the lead in making sure the president does not get his way.

Senator Coats says phone calls to his office are running 94 percent *against* lifting the ban. He expects massive disruption within the military if the measure is pushed through. Joint Chiefs of Staff Chairman Gen. Colin Powell, who opposes Clinton's plan, recently told an Annapolis midshipman, who said he was offended by the proposal, "If it strikes at the heart of your moral beliefs, you have to resign."

It is only recently that the military, reviled by many of Clinton's generation during the Vietnam era, has been restored to its former glory and to overwhelming public approval. Does the president seriously believe that the parents of the best and brightest young men and women will encourage their sons and daughters to join the military if the homosexual ban is lifted?

What about housing and tight crew quarters? What about the morale of heterosexuals and their ability to fight? What about the larger moral issue of whether homosexual practice is right or wrong?

If homosexuals are officially welcomed into the military, it won't stop there. The argument will then be that if homosexuals can fight and die for their country, the state should sanction same-sex marriages. And after that?

The pedophiles are knocking on the cultural door, asking for legitimacy. *The New York Times Book Review* recently published a favorable review of a book that praises adult-child sexual relations.

Too far out? Not when you consider how far down the moral ladder we have slipped these last thirty years. Once the standard that measures right and wrong has been removed anything becomes possible—even probable. And the decline of a moral universe brings with it an inevitable decline in the relatively humane political universe it supports.

As political scientist Glenn Tinder has written, without a moral order founded on the Judeo-Christian tradition, "The kind of political order we are used to . . . becomes indefensible."

The issue of moral decline could be used as a club to chase Democrats from control of Congress and the White House just as the soft-on-crime/soft-on-communism club beat them to death for most of the last twenty-five years.

Bill Clinton's honeymoon ended before the suitcases were unpacked. If he doesn't abandon his plan to admit homosexuals into the military, many heterosexuals who voted for him, including those morally conservative Reagan Democrats, may institute divorce proceedings.

13

AN EMBLEM OF HOPE

GEORGE GRANT

In the emotionally charged atmosphere of ideological politics, the very personal dimension of the homosexual lifestyle is often forgotten. In this poignant story, Dr. Grant relates the struggle of one man to overcome his sexual proclivities—seemingly against all odds. It probably elucidates more clearly than any argument ever could that the conflict over sexual orientation need not be an intractable conflict of unchangeable absolutes.

In ancient Greece and Rome, the festival of Bacchus was held each year to celebrate the spring harvest. It was always a chaotic and raucous affair. During the obstreperous week-long festivities, the normally sedate city states of the Pelopenese succumbed to animal passions and compulsive caprices. They profligately indulged in every form of sensual gratification imaginable—from fornication and sodomy to intoxication and gluttony. It was an orgy of promiscuous pleasure.

Bacchus was the god of wine, women, and song. To the ancients, he was the epitome of pleasure. His mythic exploits were a dominant theme in the popular art, music, and ideas of the day. Chroniclers of the age tell us that the annual carnival—called the bacchanal—commemorating his legend was actually the bawdy highlight of the year. In fact, it dominated the Helleno-Latin calendar then even more than Christmastide does ours today.

But a wild picture of immorality which only poets and mobs can understand is always simultaneously a wild picture of melancholy which only parents and emissaries can understand. Thus, even as they revelled in the streets, the ancients were troubled by a sublime sadness. It was like an ache in their throats or a knot in their stomachs but it was actually an abscess in their souls. Even as they celebrated their gay gaiety, they were forced to acknowledge their human unhappiness.

Ultimately, it was the bacchanal's smothering culture of sexual excess that proved to be the undoing of Greco-Roman dominance in the world. Historians from Thucydides and Heroditus to Himmelfarb and Schlossberg have documented that the social collapse of Greco-Romanism was rooted in the moral collapse of Greco-Romanism. The reason is as simple as it is universal.

Just as liberty and equality are opposite extremes often contrary to freedom, so sensuality and satisfaction are opposite extremes often contrary to happiness. There is in fact, no real connection between the pursuit of happiness and the pursuit of pleasure. Happiness and pleasure are indeed, in a sense, antithetical things since happiness is founded on the value of something eternal while pleasure is founded on the value of something ephemeral.

Sadly, the lessons of history are all too often lost on those who most need to learn them.

The first time I ever witnessed a "Gay Pride Parade," I was astonished—and I was reminded instantly of those ancient bacchanals. The raucous revelry, the perverse promiscuity, the orgiastic opulence, and the appolyonic abandon that I saw in the Montrose section of Houston seemed almost identical to the descriptions I had read of the Greeks and

Romans on the precipice of their demise. The malevolent scene before me could have easily been transported three thousand miles and three thousand years to the bustling Plaka under the shadow of the Acropolis without missing a beat. The sights and sounds would have been no more alien there than here.

It was on that disorienting day in that discordant place that I first encountered the despondent distress of homosexuality—it was there and then that I first met Lance.

We were as opposite as night and day. He was an exuberant participant in the day's festivities—cavorting in the aura of libertine excess. I was a member of an evangelistic team—inviting people to attend a Bible study that I conducted for those who were interested in deliverance from the captivity of that same libertine excess. He was cocky, idealistic, and tragically self-sure. I was nervous, gloomy, and profoundly self-conscious. He was only just recently "out of the closet." I was only just recently married. He was appalled by my mission. I was no less alarmed by his.

Ironically, we became quick friends.

Paradox is at the root of all true friendships. The tensions of likes and dislikes, similarities and differences, comparisons and contrasts must be delicately balanced for mere camaraderie to mature into genuine openness. And so our odd relationship grew in fits and starts.

He periodically attended our Bible study, but always left infuriated. Marshalling tattered second-hand arguments from wishful-thinking liberals and salon-chair-expositors, he railed against the very plain teaching of Scripture on the subject of sexuality in general and homosexuality in particular. He first tried the tired tack of assailing Biblical authority. Failing at that, he tried some creative exegetical gymnastics. In desperation he attacked particularly onerous episodes in Church history or obviously noxious inconsistencies in Church practice. His torturous logical contortions became convincing evidence that all too often, a man's theology is shaped by his morality—not the other way around. He simply would not give up a fancy under the shock of a fact.

Even so, Lance seriously sought for a substantive justification for his sexual orientation and practice. He had first yielded to his homosexual urges while serving a hitch in the army. In the beginning he thought that he had come to the end of his life-long search for meaning and significance. It turned out to be just another false start.

Because of the strictly enforced ban on homosexual activity in the armed services, he and his covert lovers felt more than a little inhibited. So, he opted out at the first opportunity and joined Houston's ribald homosexual community.

But even that failed to satisfy him. His unhappiness continued to gnaw at him—mind, body, and spirit. He yearned for something more than what the saturnalias of the gay bars offered. He yearned for something more than what the bacchanalias of the gay parades offered. Laboring night and day like a factory, he poured through every scrap of literature he could find on the subject—both what I gave him and what he could dredge up on his own.

"I had become desperate," he told me sometime later. "I knew that I was so lost I didn't know which way was up. I was so lonely—and the anonymous sex I had at the bars and bath houses only intensified that loneliness. The only place that I found any kind of authenticity was at the Bible study. But that grated on me terribly. I can remember sitting outside your apartment in my car debating whether or not I should go in. I felt damned if I did and damned if I didn't. Now I know that I was simply under conviction, but at the time I just knew that I was miserable."

For a few months, he tried to assuage his anguish by going to the services of a local pro-homosexual congregation. "I thought that might relieve the pressure I was feeling," he explained. "But it only made things worse. The inconsistency of that kind of pick-and-choose Christianity was obvious to me right away. I determined that I had only two choices: accept Christianity as a whole or reject it as a whole. The option of winnowing out the parts I liked and trashing the rest—cafeteria-style—just seemed like the height of hypocrisy."

Indeed, hypocrisy is the tribute which error pays to truth and inconsistency is the tribute which iniquity renders to integrity.

For several years Lance struggled with the enigmas of grace and truth. He watched as several of his former friends and lovers were alternately rescued by the Gospel or consumed by AIDS. Meanwhile, the downward spiral of his promiscuity accelerated alarmingly.

There were times when he would cut off all contact with me—for months on end. Then he would show up at my door for desperation counseling. Finally, late one Friday night, he yielded and trusted Christ for the very first time.

"There was no instant flash of revelation," he said. "No fireworks. No bells and whistles. I had just come to the end of myself."

For the longest time, he had resisted the inexorable tug of grace in his life. But all the while he knew that he did not have an opposing theory so much as a opining thirst. Ultimately that thirst drove him to drink at the sure and eternal Fountain.

Lance has been a different person ever since. "I had become convinced that I was born a homosexual. Now I know that I was just born a sinner. I never could find a cure for the former. Thankfully, the cure for the latter found me."

As a result, his life has become an emblem of hope to any and all trapped in the vicious downward spiral of licentiousness. He is happily married and the proud father of four beautiful children. "People can change. Sexual orientation is not cruelly predestined by some freak genetic code. There really is hope. I'm living proof."

CUTTING-EDGE RESOURCES

The 57 Percent Solution:
A Conservative Strategy for the Next Four Years

Where do we go from here? What should we be doing over the next four years to protect ourselves from the harmful effects of the *hole in the moral ozone?* This analysis of the 1992 election offers hope, perspective, and a balanced agenda for the silenced majority.

128 pages, paperback, $9

Unnatural Affections:
The Impuritan Ethic of Homosexuality and the Modern Church

This controversial book is a short analysis of the current crisis in the Church concerning homosexuality. It examines how the world has influenced the Church, the issue of authority, and a look at the legacy of twenty centuries of Church tradition and moral teaching.

112 pages, paperback, $8

Legislating Immorality:
The Homosexual Movement Comes Out of the Closet

This comprehensive and insightful book exposes the tragic agenda of the homosexual movement in America and why it must be reversed. It demonstrates through accurate research and analysis how the government, the media, the medical profession, the education establishment, and even the Church, have cooperated with the homosexual movement to undermine the foundations of our society.

240 pages, paperback, $10

To order by MasterCard or Visa, call toll-free *1-800-662-8629.* Or send complete payment with your order to: Legacy Communications, P.O. Box 680365, Franklin, TN 37068.